From Quantity to Number

Volume 1

From Quantity to Number

Volume 1

Mark A. Sapirie ★ *illustrations by* Marie E. Sapirie

Inlet Books

Purchase this book at www.inletbooks.com.

© 2020 Inlet Books LLC

www.inletbooks.com

All rights reserved. No part of this book may be reproduced in any form by any electronic or mechanical means (including photocopying, recording or information storage and retrieval) without permission in writing from Inlet Books.

10 9 8 7 6 5 4 3 2 1

ISBN 978-1-7351890-0-0 Inlet Books LLC

To M, T, A, P & S.

Preface

This is a first course in mathematics intended in the first instance for home school students and their parents. In fact, this book derives directly from our own home school math discussions the first year (2015–16) of which it synthesizes.

Our experience with a classical discussion-based approach to home school, a curriculum where the teacher and the student discuss together the topics of the course, is that this method is extremely effective in helping students to learn and in offering an interesting and engaging context for studies. This is as true in mathematics as it is in other subjects. And although many families have recognized the soundness of this method in their non-mathematics studies, it seems much less common for math where often families reluctantly resort to standardized commercial curricula. This book is designed to give interested families the option to pursue also in their math studies a similar discussion-based approach that can be so fruitful and enjoyable for the student and for the teacher.

A central idea to this program is the priority of forming thinking habits of mind, as opposed to drilling mechanical skills. We adhere to the principle that learning a subject should be based always in understanding and reasoning important ideas and not in rote application of a rule or a formula. So we develop and discuss concepts insofar as they offer helpful insight into understanding the topics of our studies. We focus also on how key concepts are related so that we can build up a mathematical understanding featuring coherence from which hopefully emerges for the student the tremendous beauty of mathematics and mathematical inquiry. We also emphasize the multiplicity of approaches that may exist to solve problems and to gain perspective on key ideas.

The main topic in this first set of two volumes is the development of the concept of number. We want to show the concept of number as a beautiful, rich, and evolving idea; an idea that we build up with a progressively increasing mastery of the arithmetic operations. This will lead us from number as a simple tool to help us describe quantity in the most basic sense towards the modern idea of number as the element of a continuum with a fundamental connection to core concepts underlying more advanced studies such as Calculus.

A subsidiary goal in this text is to introduce as early as possible ideas that are foundational to more advanced mathematics. We believe that these ideas are accessible if they are presented in the right way, that these ideas make discussions more interesting for students and teachers, that they will motivate students to seek out more mathematics, and that they will help future studies by providing a fertile foundation for creative thinking in mathematics. This approach also allows students to begin turning over these ideas in their minds earlier.

By way of reference, our daily math program included two parts. First a discussion on the topics set out in this text. The student participates actively in the flow of the discussion even if sometimes by indicating that a new approach to a topic is required. We kept a blank notebook handy, like a chalk board, during the discussions and we would work through many problems as we thought about the ideas that you can see in the table of contents.

The second essential part of our daily math program was an assignment for the student to work alone. Usually I provided just specifications for the assignments. So for example if you are working on multiplication, then you might indicate 20 multiplication problems say with the numbers 6 and 12 times two digit numbers between 41 and 69. In this way the teacher can tailor assignments based on the student's need while providing the student some leeway to think up problems independently. The student can learn a lot in this.

The discussions set out in this book start from beginning counting and assume no prior mathematical knowledge. Students need not even be reading yet as long as they are ready to follow along in the discussions proposed by their teacher. For the teachers, we recommend reading at least the entire chapter (preferably all the chapters) prior to beginning discussions. A preview of all the chapters will take most teachers about an hour but it will enable the teacher to see how the key ideas and themes develop and also how the examples and exercises might be modeled for additional practice if necessary. This kind of preview will help the teacher to lead discussions effectively throughout the chapters even if the teacher did not pursue math much beyond school. The chapters are not too long but they will provide the teacher with specific guidance. The text is not intended as a verbatim script but rather as a source for specific concepts, explanations, insights, examples, and exercises providing a secure framework for discussions with a clear ordering of ideas and relationships. An important aspect of our approach here is that the teacher has freedom to maneuver (which is more interesting for the teacher and the student) and also to tailor discussions to suit individual student needs.

The first focus in the text is on counting in the usual sense. Chapter one covers beginning counting. Counting up, looking towards the infinite and counting towards the infinite. We develop the arithmetic operations based on their relatedness and how they provide for counting in new ways. So after a quick overview of numbers in chapter two we turn successively to addition, multiplication, division, subtraction, and fractions. This ordering enables us to build up our work incrementally. Thus, for example, multiplication follows naturally and intuitively from addition—as repeated addition; division flows out of any multiplication expression and so is discussed immediately after multiplication; subtraction extends our understanding of number with the property of direction, and finally fractions synthesize our work on division and counting.

As we become more familiar with counting in this initial approach, we move on to consider another perspective; that of *counting closer*. In this second focus, which we broach in volume 1 and continue in volume 2 (forthcoming), we look carefully at the gaps between numbers and for which we lack a numerical representation until we come up with some ideas on how to express numbers that can describe these missing quantities. In this we use the intuitive geometric representation of the number line depicting numbers as points on the line and showing the *continuum* structure of quantities that we would like to make available in our conception of numbers also. The unbroken line is a simple way to see the goal of a continuous number structure and the difference between that continuum and the discrete collection of numbers in the rational number system.

The text provides examples and exercises. As indicated above, we further recommend assignments comprising *specifications* for problems that the student should devise and solve. The assignment specifications can be modeled on the examples and exercises while also being tailored to each student. In our experience, these student-created exercises offer a good opportunity for students to think about forming questions and provide them with the occasion to consider topics from a different perspective. Students gain insight and also the beneficial practice of active engagement through this effort.

Each student (and teacher) should work patiently with exercises until the concepts are clear and familiar. This work may entail different kinds of effort for different students depending on background. It is also a good idea to keep in mind that working through a math text is more like working through a musical score than like reading a novel. Studying math benefits from careful application and review. It is the work of an iterative process building up a structure of ideas. It is perhaps more similar to a design process than to our usual approach to reading a story book.

Throughout the text, there are several recurring themes. We have sought to identify these themes explicitly because they are interesting in themselves and they help to relate the disparate areas of study. For example, the theme of distinguishing an idea from its representation, the theme of changing perspectives, an introduction to the idea of process and of describing building numbers or computations as process, and the theme of building and finding structure in numbers. For example, the idea of building whole numbers by incrementation and the idea of building and extending our system of numbers as we discover new structure, patterns, and properties through arithmetic.

We hope that this text will encourage and help other families to include mathematical discussions in their daily study program.

Mark A. Sapirie
Tequesta, Florida
2 July 2020

Acknowledgments

Thanks to Marie for the great illustrations!

This book derives from discussions with Thomas and from our work together through many topics.

I used LaTeX to typeset the text. The formatting is based on a document style prepared by Mykel Kochenderfer and Tim Wheeler for their book *Algorithms for Optimization* and presented at JuliaCon 2019. *See* `https://github.com/sisl/tufte_algorithms_book`.

Contents

Preface vii

Acknowledgments xi

1 Beginning Counting 1
1.1 Expressing Quantity 1
1.2 How We Build Numbers 4
1.3 Counting by Groups of Ten 18
1.4 How To Write Numbers 27

2 Numbers 31
2.1 Positive Whole Numbers 31
2.2 Negative Whole Numbers 33
2.3 Rational Numbers & Other Kinds of Numbers 34

3 Addition 41
3.1 Adding One 41
3.2 Adding Twos 45
3.3 Zero 45
3.4 Order of Addition 46
3.5 Counting Groups 51
3.6 Number Patterns and Structures 56
3.7 Adding Groups of Ten 59

4	*Multiplication* 63	
4.1	*Multiplication as Repeated Addition*	63
4.2	*Multiplication as Counting Groups*	64
4.3	*Powers of Ten and Other Group Multiplications*	67
4.4	*Multiplication Patterns and Structures*	68
4.5	*The Property of Commutativity $n \cdot m = m \cdot n$*	74
4.6	*The Property of Associativity $(n \cdot m) \cdot p = (m \cdot p) \cdot n$*	75
4.7	*Building Up Answers With Easier Questions: Distributivity*	75
4.8	*Prime Factorization*	81
5	*Division* 85	
5.1	*From Multiplication To Division*	85
5.2	*Division Expresses a Counting*	87
5.3	*Longer Division*	94
5.4	*Division with Remainder*	95
5.5	*Rational Numbers*	101
5.6	*Inverse*	105
5.7	*Simplification of Rational Numbers*	106
6	*Subtraction & Directed Numbers* 109	
6.1	*Counting Down*	109
6.2	*The Other Side of Zero*	113
6.3	*Arithmetic with Subtraction & Negative Numbers*	117
7	*Fractions* 123	
7.1	*A Part of One*	123
7.2	*Unit Fractions*	124
7.3	*Whole & Fractional Quantities*	136
7.4	*Multiplying Fractions*	139
7.5	*Inverses of Fractions*	144
7.6	*Dividing Fractions*	144
7.7	*Simplifying Fractions*	151
7.8	*Adding Fractions*	154
7.9	*Proportion*	159

1 Beginning Counting

1.1 Expressing Quantity

The idea of quantity is all around us. We count things everywhere. The more that we look, the more kinds of quantities we see. We count things like apples or pears. And we will see many things to measure by counting like length, weight, time, speed, and temperature. These can all be determined as quantities. You have probably noticed that we use numbers to express quantity. And as we begin to build numbers to express quantity, let's note carefully this distinction between the two ideas

- **quantity** and

- **number**

Numbers are a *representation* of quantity. Numbers express quantities. Part of the work in understanding numbers, therefore, is to appreciate how our representation approaches what it seeks to represent. How our number system enables us to designate quantity.

Recognizing this distinction, we can identify two aspects of the work in learning to count. We must learn about quantity. And we must learn about how to express quantity. As we will see in subsequent chapters, much of the work of arithmetic can be thought of as finding different ways to re-express quantities.

Let's begin here by looking at quantity. A quantity comes up whenever we ask the questions *how much?* or *how many?*

For example, looking out the window, how many trees do I see? Figure 1.1 shows the view and we can count the trees. There is a quantity here to express and we would like to have a practical way to express it.

An obvious and convenient starting point in answering these questions is to count with our fingers. In this way we relate quantities that we observe to quantities that we know. For this we need simply to name the quantities that we express with our fingers.

Figure 1.1. Looking out the window we see three trees. We will learn how to express with numbers quantities like *three*.

Numbers are the names—the symbols—that we use to represent quantities. Our number system has been devised so that we can express big quantities fairly easily (and very small quantities too). So in this sense the number system we are learning here is practical and useful. But we ought to keep in mind nevertheless that this system is just one method among many. It is a method used by convention and by agreement and not because it is itself necessary. Our representation of quantity is separate from quantity. It is just a way to reference quantity. We could use different methods to count and in fact sometimes we will.

As a *symbol*, a number is a sign or a name that stands in the place of quantity. A number represents a quantity in the same way that a flag might represent a country or that a name might represent a person. Thus number is an example of *abstraction*, a process that is very useful in working with ideas. We can already start to appreciate this process.

Exercise 1.1. Let's say we are looking into the sea and we see eight fish. How would you write this?

Solution: You might draw a picture of eight fish. See figure 1.2. That might take a few minutes.

And what if there were many more fish?

Figure 1.2. We can draw the fish we saw in the sea.

You might also draw just one fish and then note that for what comes next a mark 'I' means 'one fish.' Then you would write 'IIIII III' to indicate how many fish you saw in the pond. This might be a bit more convenient than drawing eight fish. It saves some time. See figure 1.3.

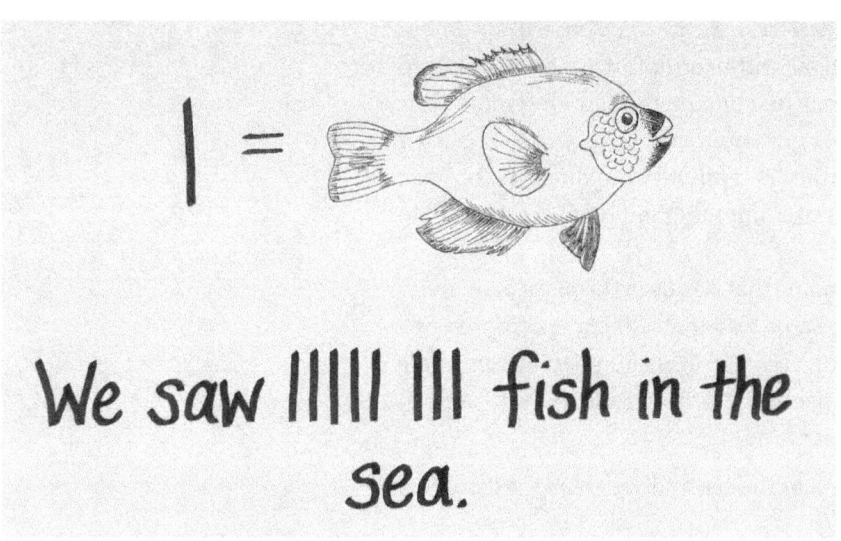

Figure 1.3. After we agree that one mark 'I' stands for one fish, we can represent the eight fish more quickly with 'IIIII III'.

Symbols are convenient for keeping track of things and working with ideas.

Exercise 1.2. Design symbols for a few animals or things around you.

Exercise 1.3. Design symbols to represent the quantities one, two, three, four, and five.

Exercise 1.4. Use your symbols to express how you would count a few things you see around you.

1.2 How We Build Numbers

If you think about what we want to do here as we get started with numbers, you could describe the challenge like this: we want to come up with a way to name quantities.

If we only needed to work with a few quantities then it would be easy. We could just choose a few different names for these quantities. But of course an interesting feature of our task is that there are very many quantities that we want to be able to name easily. And actually no matter how many quantities we name, we could always find another bigger quantity. So we want a method for naming quantities that is easy to extend; easier than just inventing a new name each time we think of a bigger quantity.

Our common agreed system for writing numbers is based on the number ten. This system provides a quick way to represent quantities however big they might be. And the key to this as we will see below is in the idea of counting groups of quantities as well as ones. When we count in this way with groups based on the number ten, we say that we are counting in *base ten*.

There are **two crucial milestones** in learning to work with numbers in our base ten system.

- First, understand the symbols for the nine numbers and the symbol for zero or nothing: 1, 2, 3, 4, 5, 6, 7, 8, 9, and 0.

- Second, understand that you can count groups of quantities as well as a things individually.

It is worth building up to these milestones patiently. We will provide the basis for a careful discussion leading to each of these milestones but each student must circle around these ideas until they are very familiar. The exercises, which may also serve as examples for more exercises that you can devise, will help you along the way in this work. Do as many as necessary.

Figure 1.4. We can count things individually and we can count groups. Here we count five apples individually. And also eggs in groups of a dozen. Here is one dozen eggs.

1.2.1 Our Ten Numerals

You must understand these ten symbols: 1, 2, 3, 4, 5, 6, 7, 8, 9, and 0. These are the written names that we use for nine numbers and for a number that we call zero and which means 'nothing.' We will see that zero is especially useful when we use a list of numerals (such as 18,028) to count groups. We refer to these ten symbols as *numerals*. They are also called *digits*, especially when they are part of a list of numerals defining a bigger number.

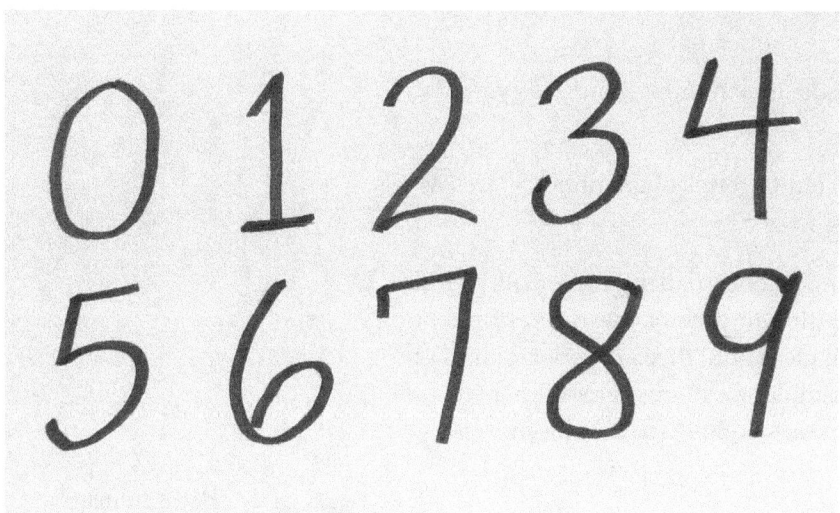

Figure 1.5. Understand these ten digits. And practice writing them too!

Exercise 1.5. For each digit shown in figure 1.5, draw the corresponding quantity of an animal or thing of your choice.

Exercise 1.6. Looking around, notice five different *kinds of things* that you can count. How many of each of those things can you see?

Solution: For example, on my desk there are books and pens and papers. There are 4 books, 2 pens, and lots of papers. We'll get back to the *lots of* papers in a moment.

Exercise 1.7. If you have a green crayon and a friend gives you a red crayon, how many crayons do you now have?

Exercise 1.8. If you have five seashells and you find one more seashell, how many seashells do you have?

Exercise 1.9. Make up and answer questions like exercise 1.8 where you add one to a quantity until you are very comfortable with adding ones to the quantities 0, 1, 2, 3, 4, 5, 6, 7, 8. You might even sometimes add 2 noticing that 2 is just two ones.

Exercise 1.10. Referring to figure 1.6, who has more whiskers: Tom or Sylvester?

Figure 1.6. Tom, the cat on the left, has five whiskers. Sylvester, the cat on the right, has eight whiskers.

Exercise 1.11. Make up and answer questions like exercise 1.10 where you observe two quantities less than ten and note which one is bigger and which one is smaller.

1.2.2 Counting Beyond Nine

There are many quantities beyond 9 that we will want to count. But we do not have a new symbol for each of these bigger quantities. We could make more symbols for bigger quantities. But even if we did that, there would always be quantities even bigger. For example, even if we designated one hundred different symbols for quantities up to ninety-nine we still might want to count quantities bigger than that. So for now let's stick with our symbols 0—9 and see if we can come up with a way to count bigger quantities nevertheless.

Exercise 1.12. Can you already think of a way to count quantities bigger than 9?

Solution: Try counting groups of ones instead of just ones.

We can use the symbols that we have to count up groups. For example, let's count the seashells in figure 1.7. But instead of counting individual shells, we will count groups of shells. How many groups of seashells can we see? We can circle them.

Once we understand the meaning of the digits from 0 through 9, the solution to exercise 1.12 is to use *groups* to count bigger quantities.

8   CHAPTER 1. BEGINNING COUNTING

Figure 1.7. How many *groups* of seashells are there?

Exercise 1.13. Count groups to describe how many birds are in the flock shown in figure 1.8.

Solution: There are several ways you can count up groups here. One way is to count groups of 2 birds. There are four groups of two birds.

The key idea to count quantities bigger than 9 is to count groups. And we count groups of 10 because we have ten symbols 0—9. Thus when we want to count the quantity that is one bigger than nine, we start the counting over, noting groups of ten. The way that we write ten indicates that we think of the quantity ten as one group of ten (and zero ones). We write a list of two digits 10. Meaning one ten. And zero ones. (*See* figure 1.9.)

Now there could be many different ways to count groups. But the way just described is the convention commonly used. It is our common practice for counting. We will count groups of ten and groups of groups of ten when we need to describe big quantities. Here we are writing out ten in English to emphasize that this refers to a name for a quantity.

There are several reasons to count groups of ten, by the way. First, ten is an easy number to work with. You have ten fingers, for example. Second, we have already learned the digits ten through 9. We can count up to 9 with a single digit.

Exercise 1.14. Can you think of a good way to express the quantity *ten* using the digits that we have learned for 0 through 9?

Solution: There are several ways to do this. See what you can come up with.

Figure 1.8. A flock of 8 birds.

To count a quantity bigger than nine, we will keep track of how many tens we count in addition to how many ones we count. Like this: the number ten is 10, meaning that we count one group of ten and nothing else. In particular, we have zero ones in addition to the one ten. One group of ten of course is made up of ten ones. The list '10' means one ten and zero ones. See figure 1.9 again.

Notice then that in considering the number 10, you can think of it in two different ways. You can think of it as 1 *of the quantity ten* or you can think of it as 10 *of the quantity one*. These two ways of thinking about the number ten refer to the same quantity. But this quantity is represented differently in thought: either as one of a quantity 10. Or as 10 of the quantity 1. There are other representations as well but this illustrates the important fact that we may represent a given quantity different ways.

Once we agree that when we have a list of two digits, the digit on the left counts tens and the digit on the right counts ones then it is convenient to write ten simply as the list one and zero: 10. So with our ten symbols, 0—9, and this simple agreement that in a list of two digits the left digit counts groups of ten and the right digit counts ones, now we can count numbers through ninety-nine which is nine tens and nine ones: 99.

If you look at some numbers around you in books and on signs, you will notice that they are made of *lists of digits*. And you already appreciate the key to understanding the meaning of these lists: you must simply know what each digit in the list counts. The method is very simple as we will see below

Figure 1.9. We write ten as a list of two digits: one and zero.

in detail: each digit counts a grouping of tens and we know exactly which grouping of ten we are counting by the position of the digit in the list.

Look at the number expressed as the list of digits in figure 1.10. This list consists in three digits. From left to right 7, 0, 8. We write 708. The digit 8 counts ones. The digit 0 count tens. And the digit 7 counts a group of ten tens. So this list of digits designates that quantity that is 7 groups of ten tens *and* 8 ones. The name for the group of ten tens is one hundred. So the number 708 counts the quantity *seven hundred and eight*. Notice that because the digit 0 is in the place counting tens, there are 0 tens to include in this quantity.

Exercise 1.15. Look again at the number 708 in figure 1.10. Can you explain why we include the zero to count tens ?

Solution: We include 0 explicitly because we have a count for hundreds: 7. The zero counting tens makes clear that the adjacent 7 to the left of the zero counts hundreds.

The first nine numbers are simple because we have enough symbols to designate each of these nine numbers individually 1 through 9: 1, 2, 3, 4, 5, 6, 7, 8, 9. And we have now just seen how to express the number for the quantity that is one more than 9. This quantity is 10.

Exercise 1.16. Think of an explanation for the idea of counting by groups.

Solution: Here is an example: an analogy to explain the idea of counting by groups. An *analogy* is like a little story that helps to explain another idea. Thinking of good (and not so good) analogies can be helpful in exploring

Figure 1.10. What is the meaning of this list of digits?

708

new concepts. Good analogies can be a lot of fun. See what kinds of vivid analogies you can come up with.

Here is an analogy. Let's say we have a big bin of golf balls that we must count. We could count the golf balls one by one. That might take a long time. But what if we had a tube that could vacuum up golf balls ten at a time. Each time we fill the tube we count ten golf balls. If we fill the tube 10 times, we have counted 100 balls. By counting the number of times we use the vacuum tube we are counting groups of golf balls.

Exercise 1.17. Using the symbols 1, 2, 3, 4, 5, 6, 7, 8, 9, and 0, form the numbers eleven through twenty where twenty means two tens.

Solution: Eleven is one more than 10. If we add one to 10 in our notation we are counting one 10 and one 1. Like this: 11. Compare this with figure 1.9.

The other numbers are similar and you should write them out carefully. Take your time with this.

Exercise 1.18. Similarly, thirty is the name for three tens. Form the numbers: thirty, forty, fifty, sixty, seventy, eighty, and ninety.

Exercise 1.19. Form the numbers: ninety through ninety-nine.

Exercise 1.20. What does the number 28 mean?

Solution: In 28, the 2 counts tens, so it indicates 2 tens; the 8 counts ones. Therefore 28 means to designate the quantity that is two tens and eight ones.

Another way to say this: 28 counts two tens and eight ones. You can practice this kind of exercise until you are perfectly familiar with the meaning of numbers up through 99. More examples are below.

Exercise 1.21. What does the number 40 mean?

Exercise 1.22. What does the number 99 mean?

Exercise 1.23. What does the number 16 mean?

Exercise 1.24. What does the number 61 mean?

Exercise 1.25. What does the number 52 mean?

Exercise 1.26. What does the number 64 mean?

Exercise 1.27. What does the number 28 mean?

Exercise 1.28. What does the number 18 mean?

Exercise 1.29. What does the number 81 mean?

Exercise 1.30. What is the number that counts nine tens and five ones?

Solution: The number 95 counts nine tens and five ones.

Exercise 1.31. What number counts three tens and seven ones?

Exercise 1.32. What number counts six tens and 1 one?

Exercise 1.33. What number counts 2 tens and 3 ones?

Exercise 1.34. What number counts 7 tens and 7 ones?

Exercise 1.35. What number counts 4 tens and 9 ones?

1.2.3 *A Note on Naming Numbers*

After you have reached the two key milestones of understanding the meaning of the digits 0—9 and also understanding how we can count groups, learning the names of the numbers is easy.

A *decade* is a whole count of the quantity 10. In referring to years, we also speak of a decade as the consecutive years with the same tens count. For example, the nineties. For the decades, notice the pattern twenty, thirty, forty, fifty showing a first digit (2, 3, 4, 5, ...) and the suffix 'ty.'

In English the naming of the teens can appear inconsistent compared to the names of the other decades. The names of the teens invert the pattern: we don't say 'ten-four' but 'fourteen.' Whereas we say 'twenty-four' to mean 24. This is just a peculiarity of language and it is different in different languages. In French, for example, you say literally ten-seven (*dix-sept*) for seventeen. Here are some exercises to help learn the names of numbers up to 100 (one hundred). You should create your own similar exercises and do as many as necessary to be perfectly comfortable naming these numbers.

Exercise 1.36. What is the name of 21?

Solution: Twenty-one.

Exercise 1.37. What is the name of 49?

Exercise 1.38. What is the name of 78?

Exercise 1.39. What is the name of 87?

Exercise 1.40. What is the name of 52?

Exercise 1.41. What is the name of 29?

Exercise 1.42. What is the name of 30?

Exercise 1.43. What is the name of 13?

Exercise 1.44. What is the name of 12?

Exercise 1.45. What is the name of 64?

Exercise 1.46. What is the name of 43?

Exercise 1.47. What is the number of this exercise?

Exercise 1.48. What is the number of the speed limit on your street?

Exercise 1.49. How many days are in this month?

Exercise 1.50. What is the name of 25?

Exercise 1.51. How many groups of 2 count 36?

Solution: To count 36 with groups of two, we need 18 groups. See figure 1.11.

Exercise 1.52. How many groups of 6 count 24?

Exercise 1.53. How many groups of 8 count 24?

Exercise 1.54. How many groups of 3 count 24?

Exercise 1.55. How many groups of 3 count 21?

Exercise 1.56. How many groups of 3 count 15?

Exercise 1.57. How many groups of 5 count 15?

Exercise 1.58. What does the name *eighteen* mean? And how do we write the number *eighteen*?

Solution: Eighteen means we are counting one ten and eight ones. We write this number 18. Notice that *teen* refers to counting one ten and *eight* refers to counting eight ones even though we write the number in the reverse order 18.

Figure 1.11. Eighteen groups of two count 36. Thirty-six cherries.

Exercise 1.59. What does the name *eighty-one* mean? And how do we write the number *eighty-one*?

Solution: **The name *eighty-one* means we are counting eight tens and one one.** We write this number 81. Notice that the *eighty* refers to counting eight tens. Contrast this with the way we count up the teens.

Exercise 1.60. For each teen quantity, write the name of the number, explain the meaning of the name (what it counts), and write the number.

Exercise 1.61. Choose ten quantities that are two-digit numbers but not teens. For each of these, write the name of the number, the meaning of the name, and the corresponding number.

1.2.4 *Adding One: a First Simple Relationship Between Numbers*

We now know how to write numbers as lists of digits. So far we have seen numbers with one or two digits. For example, the number 36 has two digits. The digit 3 tells us this number counts 3 tens and the digit 6 tells us this number also counts 6 ones. Thirty-six means 3 tens together with 6 ones.

In thinking about numbers and how to represent them, we have focused mostly on numbers as individuals and not so much in their relatedness. We can shift perspective for a moment to notice the first of many interesting connections between numbers through the idea of *adding one*. There are three benefits to doing this here. First, it will be an easy foretaste of addition.

Second, it is a first introduction to the idea of process—building rather than just describing numbers—and this is useful in computer programming where the concern is often to instruct a computer how to do and to build. And third, it will give us a good way to practice our understanding of counting and in particular to think about counting across decades.

Take the number 5, for example.

5 means five ones. In relation to the number 4, five is four ones and one more one. This illustrates a very simple *addition* of one quantity, 4, with another quantity, 1. And also the fact that the quantity 5 can be expressed as this simple addition. There is a convenient notation to write this idea of addition. The symbol '+' (plus) means add together. And the symbol '=' (the *equal sign*) lets us indicate that two expressions have the same value. We can use these symbols to write what we have just discussed

$$4 + 1 = 5$$

Figure 1.12. $4 + 1 = 5$. Illustrating the very useful + sign and = sign.

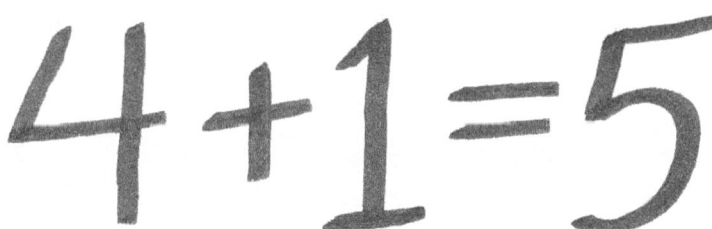

And we say *four plus one equals five* or *one added to four equals five*. The expression $4 + 1$ is on the left side of the equal sign. The expression 5 is on the right side of the equal sign. The equal sign indicates that these two expressions have the same value. Figure 1.12 illustrates the plus sign and the equal sign. Of course it doesn't matter if we add one to four or four to one as we will see later. But let's keep looking at our example here with the number five.

16   CHAPTER 1. BEGINNING COUNTING

Now we have $4 + 1 = 5$ but we could imagine one of the ones in the four shifting over to the fifth one and then we would have three plus two equals five: $3 + 2 = 5$.

$$4 + 1 = 5$$
$$3 + 2 = 5$$

Thus we have two expressions for the quantity 5.

In a similar way, we might just separate out the two ones in two like this

$$3 + 1 + 1 = 5$$

And we can continue by separating out a one from the three

$$2 + 1 + 1 + 1 = 5$$

Once more separating out the ones in the remaining two we have here

$$1 + 1 + 1 + 1 + 1 = 5$$

Here we have expressed five as a *sum* of ones only.

By the same process of adding ones we can construct any of the numbers that we have been thinking about so far. A *process* is just a way of doing or making something. The process of adding by one is called *incrementation by one*. We build a new number by adding one to a prior number. Then we repeat this with the new number. And so on indefinitely we can build numbers as big as we want.

Numbers that we can build by adding ones like this are called *whole numbers* because they can be formed by adding ones only as opposed to requiring the addition of a part of one. In contrast to whole numbers, other kinds of numbers that we will see soon cannot be formed out of ones only but also need smaller parts of the number one.

This incrementation process to build numbers is a useful perspective to keep in mind in thinking and working with numbers. It has many useful applications.

Exercise 1.62. Build the number that is one greater than each two-digit number ending in 9. For example, one such number is 49: the number that is one greater than 49 is 50. $49 + 1 = 50$.

Exercise 1.63. Let's start with 0 and repeatedly add 2. What are the first five numbers we get in this way?

Solution: 0, 2, 4, 6, 8.

Exercise 1.64. Let's start with 1 and repeatedly add 2. What are the first five numbers we get in this way?

Solution: 1, 3, 5, 7, 9.

Exercise 1.65. Let's call an *interval* a space of consecutive numbers. For example the interval from 11 to 14 consists in the numbers 11, 12, 13, and 14. We write this interval $[11, 14]$.

Count out a few different intervals of around 20 numbers. For example, count out the interval 19 through 39. Count different intervals until you are comfortable counting across each decade.

Exercise 1.66. What is $89 + 1$?

Solution:

$$\begin{aligned} 89 + 1 &= (80 + 9) + 1 \\ &= 80 + 9 + 1 \\ &= 80 + 10 \\ &= 90 \end{aligned}$$

Exercise 1.67. $79 + 1$?

Exercise 1.68. $99 + 1$?

Solution:

$$\begin{aligned} 99 + 1 &= (90 + 9) + 1 \\ &= 90 + 9 + 1 \\ &= 90 + 10 \\ &= 100 \end{aligned}$$

Exercise 1.69. $69 + 1$?

Exercise 1.70. $49 + 1$?

Exercise 1.71. $39 + 1$?

Exercise 1.72. $19 + 1$?

Exercise 1.73. $20 + 1$?

Exercise 1.74. $90 + 1$?

1.2.5 Roman Numerals

One way to appreciate the distinction between quantity and number is to learn another counting system. For example, the Romans did not use the number symbols we use. They used Roman numerals such as: I for 1 and X for 10.

Exercise 1.75. Using the Roman numerals I and X, write out the number thirty-three, that is 3 tens and 3 ones. Write it also using our numbers.

Exercise 1.76. Choose a few two digit numbers less than 40 and write down the corresponding Roman numeral.

Exercise 1.77. Optional. Research the other symbols that the Romans used to express quantity and use what you find to express quantities in our number system and also in the Roman numeral system.

1.3 Counting by Groups of Ten

Let's now return to a crucial idea in our number system: we represent numbers by counting groups based on the number ten. We call these groups also the *powers of ten* for reasons that we will see later. For now, the important idea is to understand the pattern for building these groups of 10 and it is very simple.

Consider a bigger number.

1,215. This number corresponds to the following quantity

1 thousand and 2 hundreds, 1 ten, and 5 ones.

We can write

$$1,215 = 1 \text{ thousand } + 2 \text{ hundreds } + 1 \text{ ten } + 5 \text{ ones}.$$

We read this number *one thousand two hundred and fifteen*.
In the same way

$$12 = 1 \text{ ten } + 2 \text{ ones}.$$

Or

$$2,001 = 2 \text{ thousands } + 0 \text{ hundreds } + 0 \text{ tens } + 1 \text{ one}.$$

Example 1.1. Understanding the meaning of a bigger number, 1,215.

1.3.1 Powers of Ten

For reasons that we will see later the groups of ten counted by digits in a number are called *powers of ten*. Each digit in a number counts a different power of ten. We will use these ideas to build up our number counting system by counting up from one. See figure 1.13 for some powers of ten.

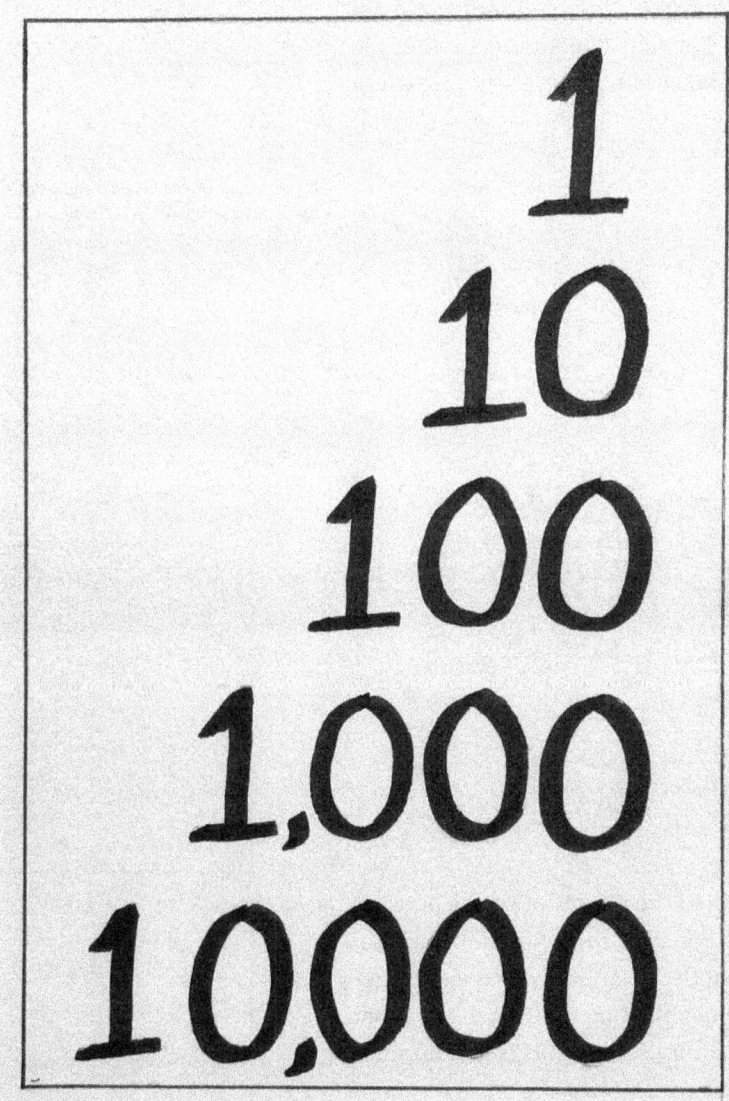

Figure 1.13. The powers of ten are the key to understanding long numbers.

There is a convenient notation for the powers of ten that is worth learning early because it makes work a lot easier. The first power of ten (also described as 10 *to the power of* 1) is 10. We say ten to the power of one is 10. We write

$$10^1 = 10$$

10 *to the power of* 2 means the group ten tens. That is, we take a group of ten, ten times and add those together. Be careful to note that we are not here speaking of merely two groups of ten. Two tens of course expresses the quantity 20. Ten to the power of two, on the other hand, means that we count 10 tens.

Figure 1.14. We can count the quantity 100 as ten groups of ten.

So think of ten as one kind of group. Ten to the power of two is another kind of group. It includes ten groups of ten. See figure 1.14 for an illustration of the quantity 10 groups of ten. If you count this quantity with ones (rather than by tens), then we say that we have the quantity *one hundred*. Notice that one hundred has the same *value* as the quantity ten tens. It is just another name for the same quantity. In power notation we write:

$$10^2 = 100.$$

In a subsequent chapter we will see a convenient notation for the expression ten tens thus

$$\text{ten tens} = 10 \times 10$$

Or also that

$$\text{ten tens} = 10 \cdot 10$$

With this notation we can re-express the ten to the power of two

$$10^2 = 10 \times 10 = 100$$

And also

$$10^2 = 10 \cdot 10 = 100$$

The symbols '·' and '×' mean *times*. $10 \cdot 10$ means that we count the quantity 10, ten times.

Ten to the power of three means we count ten to the power of two ten times. That is, take ten copies of one hundred. Or count out one hundred, ten times. We write

$$10^3 = 10 \cdot 100 = 1,000$$

Or also

$$\text{ten times ten tens} = 1,000$$

And

$$\text{ten hundreds} = 1,000$$

Or also

$$1,000 = \text{one thousand}$$

Figure 1.15 illustrates three different ways to think of counting one thousand and the corresponding numbers. One thousand. Ten hundreds. One hundred tens.

The ordinary name for this number is *one thousand*.

We can go on in this way building the powers of ten and we will do that in the section below.

Exercise 1.78. Describe the meaning of the quantity expressed when we say *we have counted three fours* in contrast with the meaning of the quantity expressed by the number 444.

Solution: When we count three four we are counting three groups of four. This means we refer to the quantity $4 + 4 + 4 = 12$. In contrast when we count the number 444, we are counting four hundreds, four tens, and four ones. That is, $400 + 40 + 4$. The difference in these descriptions is that in the first case we are counting three groups of four. Whereas in the second description, through the agreement of our number system, we are counting four hundreds, four tens, and four ones.

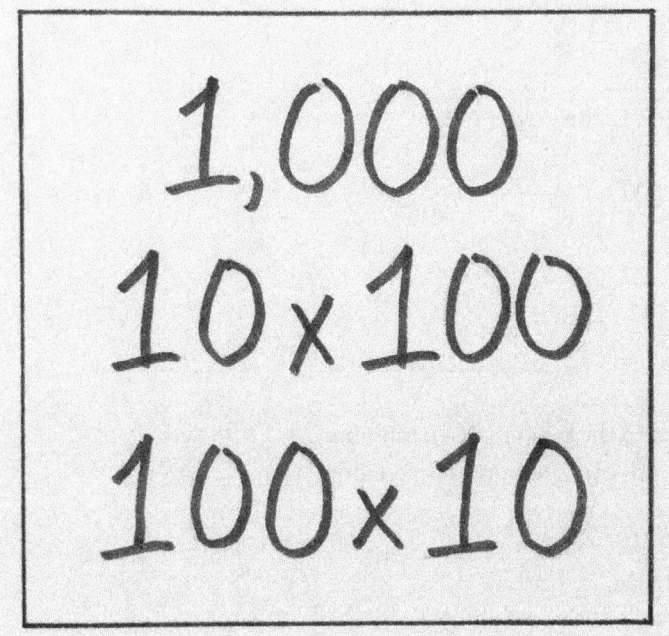

Figure 1.15. Three ways to count one thousand. One thousand. Ten hundreds. One hundred tens.

Exercise 1.79. What is the biggest quantity that we can represent with a three digit number?

Solution: The biggest single digit that we have is 9. Therefore the biggest quantity that we can express with three digits is 999. This counts nine hundreds, nine tens, and nine ones. Its name is *nine hundred ninety-nine*. It is one less than one thousand.

Exercise 1.80. What is the smallest quantity that we can express with three digits?

Solution: Zero is the smallest quantity we have expressed in a single digit. Therefore the smallest quantity that we can express with a three digit number is 100. The quantity that is one less than this is expressed with two digits: 99.

Exercise 1.81. What does the following number mean? Say also its name: 56.

Solution: 56 means to count five tens and also six ones. Its name is *fifty-six*.

Exercise 1.82. What does the following number mean? Say also its name: 506.

Solution: 506 means to count five hundreds and also six ones. Its name is *five hundred six*.

Exercise 1.83. What does the following number mean? Say also its name: 560.

Solution: 560 means to count five hundreds six tens and zero ones. Its name is *five hundred sixty.*

Exercise 1.84. What does the following number mean? Say also its name: 516.

Solution: 516 means to count five hundreds one ten and six ones. Its name is *five hundred sixteen.*

Continue thinking about numbers in this way until you are very comfortable with the meaning and names of two and three digit numbers.

1.3.2 A Few Powers of Ten

Here are the first seven groups that the digits count in a number of our base ten counting system. The first seven powers of ten. With these powers of ten we will be able to write numbers into the millions. The important point here is to understand that a number written as a list of digits represents a sum of the counts of these powers of ten. We will discuss powers in detail later but in case you are wondering about powers of ten, let's have a quick look.

$$10^0 = 1$$
$$10^1 = 10$$
$$10^2 = 10 \cdot 10 = 100$$
$$10^3 = 100 \cdot 10 = 1,000$$
$$10^4 = 1,000 \cdot 10 = 10,000$$
$$10^5 = 10,000 \cdot 10 = 100,000$$
$$10^6 = 100,000 \cdot 10 = 1,000,000$$

So each successive power of ten increases the size of the prior power of ten by counting that prior group ten times. This means we count out the prior power of ten ten times to get the next power of ten. We say that each power of 10 is a *factor* of 10 greater than the preceding power of ten. Sometimes when we are using powers of ten to get a sense of the size or scale of things, we speak of the powers of ten as *orders of magnitude*. Being attuned to orders of magnitude is often useful in thinking quickly about new situations and scales. It is a good idea to practice this.

There is, by the way, an excellent video from the Eames Office called *The Powers of Ten* (1977) available on the internet illustrating the orders of magnitude and the powers of ten that we experience in our universe from the very large scale to the very small scale. It is easy to look up and we recommend watching it.

Second, notice that the power number (the *exponent*) in the power of ten indicates the number of zeros following the 1 in the corresponding number. For example $10^3 = 1,000$. The number 3 is the exponent in the power of ten and $1,000$ has three zeros. The number 10 in 10^n is called the *base*.

Notice finally that although we have names for the powers of ten, for example 10^3 is $1,000$ and called 'one thousand,' you might find it convenient sometimes to think of the same number in different ways. For example, you might think of $1,000$ as ten hundreds.

Figure 1.16 illustrates powers of ten in terms of distances that a student might relate to.

Exercise 1.85. Make your own table similar to the one in figure 1.16 with distances of your choice.

	Powers of Ten	Orders of Magnitude (meters)
1	10^0 one	table height (1 m)
10	10^1 ten	trees (10 m)
100	10^2 hundred	baseball field (100 m)
1,000	10^3 thousand	to library (1,000 m = 1 km)
10,000	10^4 ten thousand	to Juno Beach (10 km)
100,000	10^5 hundred thousand	to Miami (100 km)
1,000,000	10^6 million	to DC (1,000 km)
10,000,000	10^7 ten million	to Paris (10,000 km)
100,000,000	10^8 hundred million	to Moon ($3 \cdot 100,000$ km)
1,000,000,000	10^9 billion	(1 million km)
10,000,000,000	10^{10} ten billion	(10 million km)
100,000,000,000	10^{11} hundred billion	to Sun (150 million km)

Figure 1.16. Reference distance orders of magnitude illustrating the powers of ten.

1.3.3 Multiplication by 10 (or by powers of 10)

If you are comfortable discussing multiplication by ten at this point here is a useful detail. We will develop multiplication subsequently so it is not a problem to skip this discussion.

In multiplication by 10, notice that any number n multiplied by 10 equals the number n with a 0 appended. For example $4 \cdot 10$ is 40.

Why is this? Because we are counting 4 ten times. Likewise, more generally we might take n ten times. So each digit in the number, prior to multiplication by ten, is ten times greater. In other words, if we look at the number n prior to multiplication by ten, each digit gets shifted to the left one place as a result of the multiplication. Each digit is counting ten times more than it was prior to the multiplication. Figure 1.17 illustrates multiplication by 10.

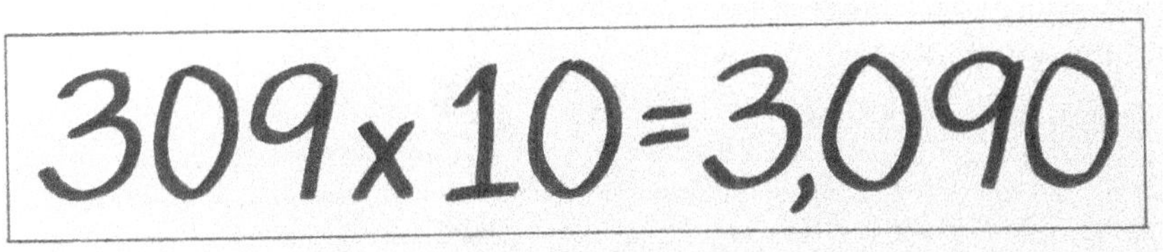

$$309 \times 10 = 3,090$$

Figure 1.17. Multiplying 309 by 10 to illustrate the effect of multiplication by 10.

What is $10,000 \times 10,000$?

$$10,000 = 10^4$$

The number in the exponent also indicates the number of zeros following the one on the left. Think back to the definition of powers to see this.

If we multiply a powers of ten that are expressed in this form, we can simply add the exponents.

$$10^4 \times 10^4 = 10 \cdot 10 \cdot 10 \cdot 10 \times 10 \cdot 10 \cdot 10 \cdot 10$$
$$= 10^8$$
$$10^4 \times 10^4 = 10^{4+4} = 10^8$$
$$10,000 \times 10,000 = 100,000,000$$

Do you see why this works? We will look at it in detail later but it is not a bad thing to think about already if you would like. Start with the meaning of a power of 10 as indicated above. What does 10^4 mean?

Example 1.2. The powers of ten make it very easy to express very big numbers.

1.3.4 An Optional Discussion on 10^{n+1}

Each next higher power of ten is simply the prior power of ten counted ten times. Figure 1.18 shows how we build the next higher power, 10^{n+1}, given the nth power of ten. The letter n here stands in the place of a number to indicate that you can think of n as any whole number. This expression tells us that if we know the nth power of 10, then we can determine the $(n+1)^{st}$ power also. We will return to this notation in detail in a later chapter. By the way, this use of a letter to stand for numbers, and to name numbers, is a useful practice to know. It makes it easier to think about ideas on numbers.

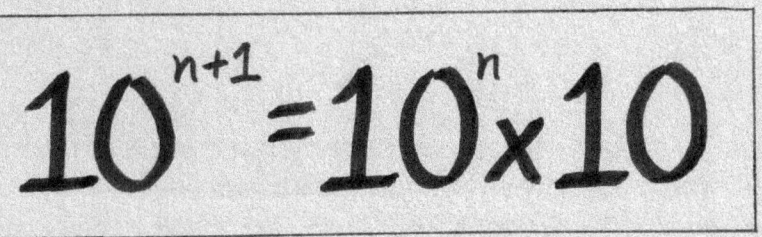

Figure 1.18. Ten to the power of $(n+1)$ in terms of ten to the power of n.

1.3.5 An Optional Little Discussion: 10^0

We have discussed building up groups of numbers—the powers of ten—beginning with 10^1. But what about 10^0?

$10^0 = 1$. Why?

At first we might think that 10^0 would equal 0. But this is not correct. To see this, let's pretend that 10^0 *does* equal 0 and think about what this would mean. We will show this would lead us to a statement that we know is wrong. So the assumption must be false. This is a *proof by contradiction*. Here is the reasoning.

Assume that
$$10^0 = 0$$

We know also that

$$10 = 10^1$$

We can write this as

$$10^{0+1}$$

From the definition of a power of ten featured above (we will also look at this property in more detail subsequently):

$$10^{0+1} = 10^0 \cdot 10^1$$

Now, using our assumption that $10^0 = 0$ we would get:

$$10 = 0 \cdot 10$$

i.e., $10 = 0$.

This is absurd. So our assumption must be false. □

The proof above shows us what 10^0 is not, but we still have to find the right answer. In fact 10^0 must equal 1. This property follows from the nature of powers arithmetic (which we will see in detail later but for completeness in the discussion, here is a quick look). We can figure it out by thinking about what it must be so that our counting by powers of ten is consistent.

For example

$$10^2 = 10^{2+0} = 10^2 \cdot 10^0 = 100 \cdot 10^0$$

So

$$10^0 \text{ must equal } 1.$$

1.4 How To Write Numbers

With the powers of ten, we are ready to count bigger numbers more generally. In our system, a number is a list of digits. Each digit tells you a count for a power of ten beginning on the right with a count for the ones–the zeroth power of ten. Here are some examples.

Exercise 1.86. What does 11 mean?

Solution: 11 means you have 1 ten and 1 one: $11 = 1 \cdot 10 + 1 \cdot 1$.

Exercise 1.87. What does 23 mean?

Solution: 23 means you have 2 tens and 3 ones: $23 = 2 \cdot 10 + 3 \cdot 1 = 20 + 3$.

28 CHAPTER 1. BEGINNING COUNTING

Exercise 1.88. What does 7 mean?

Solution: 7 means a count of 7 ones: $7 \cdot 1$.

Exercise 1.89. What does 483 mean?

Solution: 483 means a count of 4 hundreds and 8 tens and 3 ones: $483 = 4 \cdot 100 + 8 \cdot 10 + 3 \cdot 1 = 400 + 80 + 3$.

Exercise 1.90. What does $2,008$ mean?

Solution: $2,008$ means a count of 2 thousands, 0 hundreds, 0 tens, and 8 ones: $2,008 = 2 \cdot 1,000 + 0 \cdot 100 + 0 \cdot 10 + 8 \cdot 1 = 2,000 + 8$.

Exercise 1.91. What does 56 mean?

Solution: 5 tens and 6 ones.

Exercise 1.92. Similarly, interpret 756.

Exercise 1.93. What is the meaning of 10^3 ?

Solution: It means $10 \cdot 10 \cdot 10$. This is the number *one thousand*: $1,000$.

Exercise 1.94. What does 33 mean?

Exercise 1.95. What does $7,921$ mean?

Exercise 1.96. What does $7,901$ mean?

Exercise 1.97. How do you write the number *one thousand and four*? What does each digit mean?

Solution: $1,004$. This number counts 1 thousand and 4 ones. The two zeros in the middle count 0 hundreds and 0 tens.

Exercise 1.98. What about $47,921$? What does that mean?

1.4.1 Right to Left

Why do we build numbers up from right to left—as in the ones are counted in the first digit on the right and the digits to the left count bigger powers of ten—whereas we read left to right? The answer is that our numbers come from the Hindu-Arabic counting system and in Arabic one reads from right to left. This format has the benefit in English of indicating the largest power of ten first (*i.e.*, to the left) as we read.

It is important to remember that there are many different ways to write the same number even in our number system. In particular, as you learn to count you should remember that every number is equal to a sum of ones. For example, $4 = 1 + 1 + 1 + 1$. In this way as soon as you know how to count you actually also know how to add numbers. We will go more into addition later. In the meantime think about different ways you can express the same number.

Exercise 1.99. Read the following numbers and detail what each digit in each number counts:

$4,235$. 363. $10,001$. $99,999$. 101. 72. $57,908$. $507,900$. $1,000,000$. $707,002$.

1.4.2 Next Steps

Work through the exercises above and similar exercises that you make up so that you become very comfortable with the ideas in this chapter. Take your time with this. Here are two points to start with. First, in learning the number names, focus on meaning. What does a given number mean? What quantities are the digits counting? Second, note that in learning the names of two digit numbers, the naming for the teens is reversed when compared to the decades. Gradually practice building two digit numbers from their meaning.

By prioritizing understanding the meaning of numbers, it will be easier for students to learn to figure out a new number even without knowing the name. The milestones in this are: 1) understanding the symbols 0—9 and then 2) counting by groups. In learning the names be sure to check that you can go from the name to the meaning and also from a description of a quantity to its name. This has been set out in exercises above.

To learn the names of the two digit numbers it might be helpful to proceed in two steps. Begin practicing just the decades. Then consider the two digit numbers generally. To gain familiarity in counting up to 100, it is helpful to know the idea of adding one and how this works to cross from one decade into the next. For example, to count from 48 to 52.

2 Numbers

We have just considered how to count and to write numbers. We will soon start learning addition. But before we get to addition, let's consider for a moment the idea of numbers more generally.

This chapter is a brief supplemental discussion and a bit of a preview while we continue to practice counting. The purpose here is to develop the idea of how we are constructing numbers. The work of building numbers turns out to be intricate as we will see. This little chapter also provides the opportunity to become familiar with some more useful notation and number concepts.

2.1 Positive Whole Numbers

In building up our system of counting whole numbers, we learned not only how to count but actually also how to add numbers—the digits of a number count powers of ten and the meaning of the number is in the sum of these powers of ten. Thus when we write the number 18, we mean one ten and eight ones.

The numbers that we have seen so far arise naturally in our experience every day. We count things all the time. The number system that we have built might be thought of as a solution to the problem: what is a good way to keep track of things that we encounter in our lives?

Moreover, with two simple ideas we saw that we can build all the numbers that we need to count in our work so far:

- **the number one** and

- **adding one to another number**

    In this way, we can count

$$2 = 1 + 1$$
$$3 = 2 + 1$$
$$4 = 3 + 1$$
$$5 = 4 + 1$$

and so on indefinitely. These numbers are the *positive whole numbers* or *positive integers*. That they are positive means they are greater than zero as opposed to *negative* numbers, which are less than zero. And they are *integers* or *whole numbers* because they may be constructed by adding ones and without requiring a fractional part of one.

As we work with positive whole numbers and basic arithmetic operations (addition, subtraction, multiplication, division, powers, and roots) that arise naturally with them, we will encounter new problems that we cannot solve if we confine ourselves to the positive whole numbers alone.

Let's consider an example. We understand now that

$$3 + 2 = 5$$

We can think of 5 as $3 + 2$. So now when someone asks us to take two away from five, that makes sense. We see that the answer is 3. We write five, take away two, equals three. The symbol '-' expresses the idea of taking away a quantity from another. That is, subtraction. Here we subtract 2 from 5 and we write

$$5 - 2 = 3$$

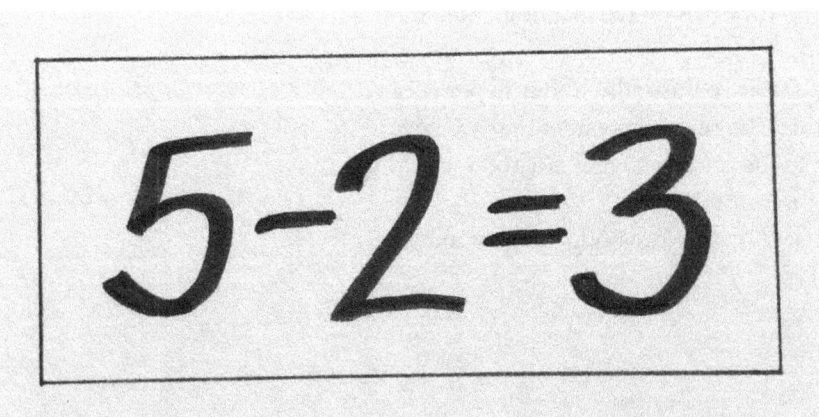

Figure 2.1. Subtracting two from five.

What if someone says next, *what is 2 − 5? How do we express this?* The new idea here is that we want to take away a bigger quantity from a smaller quantity. Can we do that? And how would we express that idea?

$$2 - 5 = ?$$

This question leads us to extend our understanding of number. To find the answer here we need a new kind of number—the *negative whole numbers*.

2.2 Negative Whole Numbers

We can build negative whole numbers (also called *negative integers*) just as we built up positive whole numbers. Starting at zero, we take away one. What does this yield? It is very similar to addition except that we seem to be moving in the *opposite direction*. We will indicate the opposite direction with the '-' sign.

$$0 - 1 = -1$$

Figure 2.2. With negative numbers, we obtain another useful property for numbers: direction.

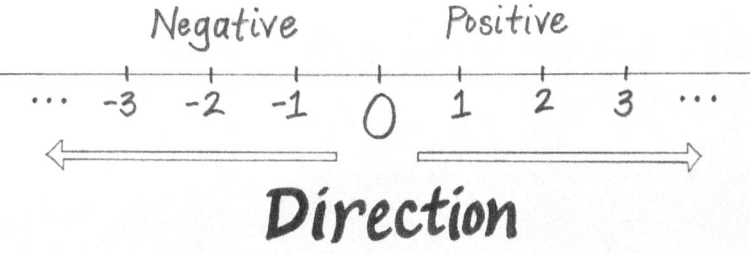

Figure 2.2 illustrates that the minus sign indicates the idea of direction in counting down.

Just as we continued the process with addition (incrementation), we can do the same with subtracting 1.

$$-1 - 1 = -2$$
$$-2 - 1 = -3$$
$$-3 - 1 = -4$$

Repeating this subtraction by 1 lets us reach each negative integer. Now we can answer questions like

$$2 - 5 = ?$$

By the same process that we just saw, we can think of $2 - 5$ as

$$2 - 5 = 2 - (1 + 1 + 1 + 1 + 1)$$

We then successively subtract each 1.
In this way we find

$$2 - 5 = -3$$

Notice that we used *parentheses* to group the ones of the five

$$2 - 5 = 2 - (1 + 1 + 1 + 1 + 1)$$

Using parentheses to show structure like this is common and very convenient. We will discuss this also later.

Exercise 2.1. Compute $2 - 6$, $2 - 7$, and $2 - 8$.

Exercise 2.2. Compute $10 - 10$ and $10 - 11$.

Exercise 2.3. Compute $8 - 0$ and $0 - 8$.

Exercise 2.4. Can you figure out what is $-1 - (-1)$? We will cover this carefully subsequently.

Exercise 2.5. Compute $-5 + 5$ and $-5 - 5$.

Exercise 2.6. Compute $19 - 20$ and $20 - 19$.

2.3 Rational Numbers & Other Kinds of Numbers

If this is your first read through the topics we have seen so far you can skip to the next chapter. But if you are a little familiar with these concepts, the discussion in this section might provide a helpful context for the following discussions.

One way to think about new kinds of numbers is that they are created to help us find answers to new kinds of questions. For example, in the discussion above we extended our understanding of whole numbers also to encompass negative whole numbers. These let us track values that diminish below zero and the concept of negative number brings to our understanding of number a sense of direction: positive and negative.

To show further this idea of building up numbers we will look here briefly at three more new kinds of numbers: *rationals, irrationals,* and *complex numbers.* We will not go into details at this point. Our purpose now is only to show how new kinds of numbers have emerged to help us answer more questions. And hopefully to inspire.

In counting groups we have seen expressions like five twos

$$2+2+2+2+2$$

The multiplication notation we introduced provides a more concise expression

$$2+2+2+2+2 = 5 \cdot 2 = 10$$

Or

$$5 \cdot 2 = 2+2+2+2+2 = 10$$

Let's look carefully at this expression $5 \cdot 2 = 10$.

The dot ($\cdot$) between the five and the two means take two five times. This is just a matter of convenient notation. We will see much more of this later.

But for now let's focus on another way to think about this expression. It tells us that five goes into ten twice. It points to a relationship between the numbers 5, 2, and 10 that can be expressed differently starting with the ten and counting how many fives are in ten. The answer is two. We can express this directly as

$$\frac{10}{5} = 2$$

Or

$$10 \div 5 = 2$$

We say ten divided by five equals two. The symbols '$\div$,' and the fractional representation ($\frac{10}{5}$) all mean the same thing. They mean that we are interested in how many of the second number we need to count the first number. Here this means how many fives must we count to get 10. Figure 2.3 illustrates the meaning of this division notation.

Figure 2.3. We can think of division as a counting. For example how many pairs of penguins are on the iceberg? Ten divided by two expresses this idea: how many twos must we count to get ten? Five!

Consider

$$\frac{10}{2} = 5$$

The expression on the right hand side of the equal sign means how many twos must we count to get 10? The answer is 5. This is another perspective on the relationship expressed in the multiplication $5 \cdot 2 = 10$.

But what if we ask instead how many threes we must count to get ten?

$$\frac{10}{3} = ?$$

If we start counting threes, we have two threes for six, three threes for nine, four threes are twelve. Three doesn't go into ten a whole number of times. You would need to count three threes and then just one out of the fourth three to get 10.

$$3 \cdot 3 = 9$$
$$4 \cdot 3 = 12$$

Here to count 10 we really need

$$3 \cdot 3 + 1$$

The number 1 is just one third of 3 for we have $1 + 1 + 1 = 3$. That is, 1 is the part of 3 such that three of these parts make up the whole. This satisfies the definition of one third which we can also indicate: $\frac{1}{3} + \frac{1}{3} + \frac{1}{3} = 1$. Figure 2.4 illustrates how three thirds make up 1.

And there is nothing to stop us from considering another extension of the numbers that we know. We will include not only whole numbers but also all the numbers that we can form by dividing one whole number by another. This includes the number $\frac{1}{3}$, the number obtained by dividing the whole number one by the whole number 3. This new expansion of numbers forms the *rational numbers*. These are numbers that may be expressed as a *ratio* (division) of two whole numbers. It turns out that the rules of arithmetic for whole numbers also work for rational numbers. But we will look at this in detail later. For now the point is simply that we can build up new numbers to answer more questions.

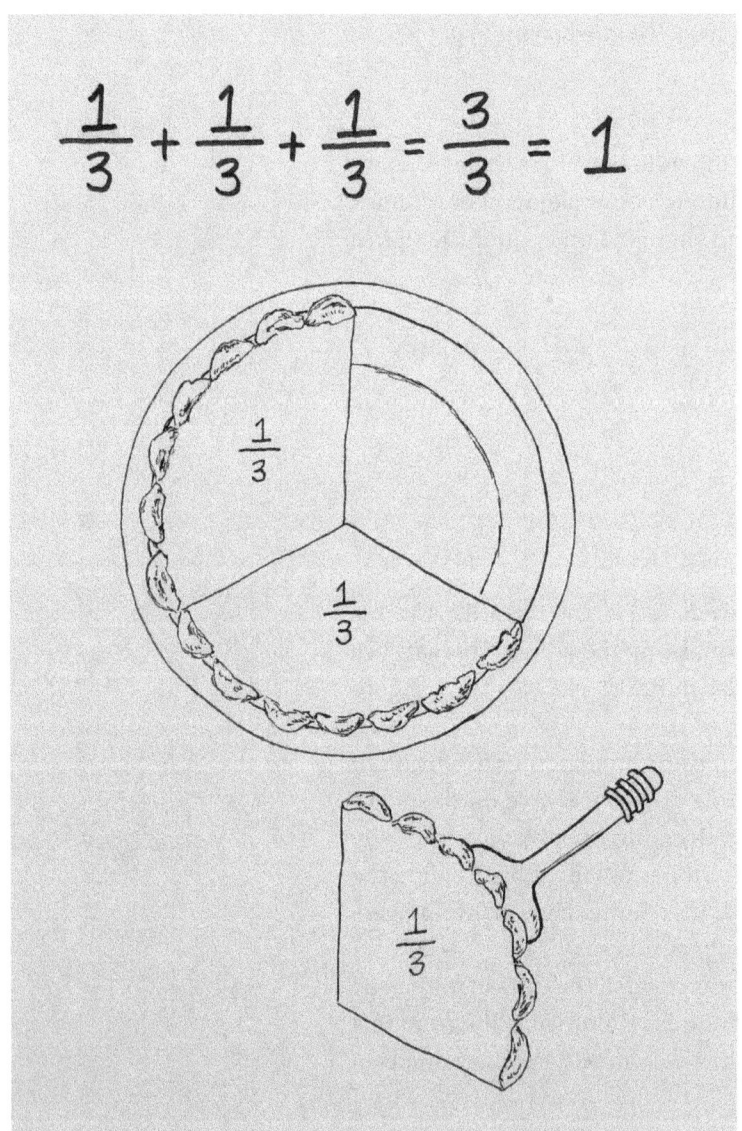

Figure 2.4. Showing how three thirds make one whole.

The two other types of numbers that we will discuss here briefly are *irrational* and *complex numbers*. Irrational numbers arise when we consider questions like what number times itself equals two?

We can think of this in the following terms. What is the number, x, so that

$$x \cdot x = 2$$

The number that we seek is called the *square root of two* and is written

$$x = \sqrt{2}$$

We will see that

$$\sqrt{2} \cdot \sqrt{2} = 2$$

But it turns out *we cannot express x as a ratio of two whole numbers*. There are no whole numbers a, b so that $\frac{a}{b} = \sqrt{2}$. x is not a rational number.

To get a sense of what is the $\sqrt{2}$ we will look closely at the idea of continuity of points on a line and how to characterize the process of approaching a specific point on a line. For now simply be aware that there are numbers like $\sqrt{2}$ that we call *irrational numbers* and that these numbers cannot be expressed as rational numbers.

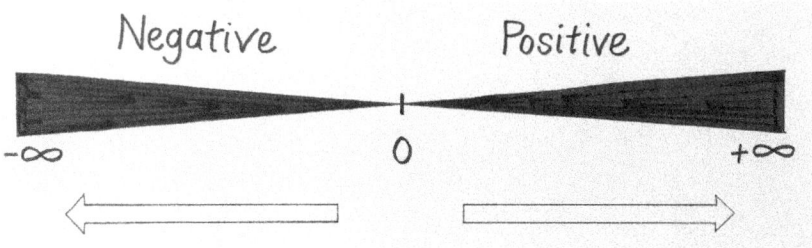

Figure 2.5. The real number line. Including the characteristic of direction achieved with negative numbers.

When we start to think of numbers as corresponding to the points on a continuous line (this is very useful as we will see) then we will need the irrational numbers to fill in gaps that would otherwise exist in our number line if we had only rational numbers to work with. The rational numbers and the irrational numbers together constitute what are called the *real numbers*, the numbers that we can use to set up a complete correspondence between numbers and points on a line. Figure 2.5 represents the real numbers as the points making up a geometric line.

Complex numbers can be seen to arise similarly when we ask the question what is the square root of -1 ($\sqrt{-1}$)? That is, what is the number k for the equation $k \cdot k = -1$. To answer this question, we invoke a new kind of number named an *imaginary number*: $i = \sqrt{-1}$. Complex numbers then are all the numbers that we can make by adding $a + i \cdot b$ where a and b are any real numbers. That is, complex numbers are numbers of the form $a + i \cdot b$ for any real numbers a and b, and $i = \sqrt{-1}$.

3 *Addition*

In our prior discussions we looked at counting. Once we know how to count we actually also know how to add. In this chapter, we will develop addition. Very likely you already understand intuitively how to add. A little practice will help you to become familiar with some useful ideas.

3.1 *Adding One*

Recall that we built up our numbers starting with the number one and adding ones. In this way we built 2, 3, 4, 5, and so on. Figure 3.1 represents a number-making machine based this idea of incrementation, addition by one.

Figure 3.1. An imaginary mathematical machine that makes positive whole numbers. Call it an incrementer.

From this construction we see that each whole number is one plus its preceding number. We express this idea generally

$$\text{next number} = \text{prior number} + 1$$

So for example we have

$$8 = 7 + 1$$

and also

$$9 + 1 = 10$$

From our construction we also know that each whole number can be expressed as a sum of ones. For example

$$5 = 1 + 1 + 1 + 1 + 1$$

With this observation we should always be able to figure out at least one way to add two numbers. Even if the numbers that we are adding are big, we can always break one of them down into a sum of ones and then simply count up those ones starting from the other number. In this way we join the second quantity to the first.

$$9 + 3 =$$
$$9 + 1 + 1 + 1 =$$
$$(9 + 1) + 1 + 1 =$$
$$10 + 1 + 1 =$$
$$(10 + 1) + 1 =$$
$$11 + 1 = 12$$

Example 3.1. Adding three to nine is easy when you understand the structure of three. $3 = 1 + 1 + 1$.

Thus to find the answer to nine plus three we need only know that three means three ones or one plus one plus one and then to count up three ones starting from nine.

With this idea we might even imagine an adding machine that first takes apart into a sum of only ones, one of the numbers in the addition and then increments the other number with each of the ones.

3.1.1 A Note on Parentheses

When we write a sequence of arithmetic operations it is sometimes convenient to use *parentheses* to group numbers. This shows the order of the steps we take as indicated above, or to show structure in an expression.

Previously we saw the symbol '=', the equal sign. It means to say that the *value* of the expression on the left hand side (LHS) of the '=' is the same as the value of the expression on its right hand side (RHS). An *expression* is like a math sentence and it contains numbers and operations like addition. Of course you could write the LHS and RHS expressions on either side of the equal sign.

Exercise 3.1.
$$7 + 3 =$$

Exercise 3.2.
$$6 + 4 =$$

Exercise 3.3.
$$9 + 1 =$$

Exercise 3.4.
$$8 + 2 =$$

Exercise 3.5.
$$8 + 5 =$$

Exercise 3.6.
$$8 + 10 =$$

Exercise 3.7.
$$7 + 8 =$$

Exercise 3.8.
$$5 + 3 =$$

Exercise 3.9.
$$9 + 6 =$$

Exercise 3.10.
$$3 + 6 =$$

Exercise 3.11.
$$2 + 19 =$$

Exercise 3.12. Choose pairs of numbers and add them together until you are comfortable with simple additions of single digit numbers.

Exercise 3.13. Choose a number and note all the ways that you can express that number with addition.

Solution:

$$\begin{aligned}
8 &= 7+1 \\
&= 6+2 \\
&= 5+3 \\
&= 4+4 \\
&= 3+5 \\
&= 2+6 \\
&= 1+7
\end{aligned}$$

Notice here again that the order of the numbers you are adding does not change the answer. $7+1 = 1+7 = 8$. This makes sense. It doesn't matter if you start with one and then add seven or if you start with seven and then add one. You end up with eight. Another way to think about this is if you have eight in your hand you can think of it as two groups—a group of one and a group of seven. You can add these together either way: seven first or one first. You have eight in your hand either way.

Sometimes it might be helpful to think of the number eight as one of the possible simple addition expressions. For example, if you want to compute $7 + 8$, you might think

$$\begin{aligned}
7 + 8 &= \\
7 + (3 + 5) &= \\
7 + 3 + 5 &= \\
(7 + 3) + 5 &= 10 + 5
\end{aligned}$$

We recognize the last expression as the definition of 15 and you know therefore

$$7 + 8 = 10 + 5 = 15$$

3.2 Adding Twos

In the last section we saw that we can easily add any two numbers simply by breaking the second number into a sum of ones and then adding the ones incrementally to the first number. If you are comfortable with this idea then you will notice a quicker way to do the addition: instead of breaking the second number into a sum of ones you can break it into bigger numbers, say twos or threes or fours.

$$9 + 13 =$$
$$9 + 2 + 2 + 2 + 2 + 2 + 2 + 1 =$$
$$(9 + 2) + 2 + 2 + 2 + 2 + 2 + 1 =$$
$$11 + 2 + 2 + 2 + 2 + 2 + 1 =$$
$$(11 + 2) + 2 + 2 + 2 + 2 + 1 =$$
$$13 + 2 + 2 + 2 + 2 + 1 =$$
$$(13 + 2) + 2 + 2 + 2 + 1 =$$
$$(15 + 2) + 2 + 2 + 1 =$$
$$(17 + 2) + 2 + 1 =$$
$$(19 + 2) + 1 =$$
$$21 + 1 = 22$$

Example 3.2. Adding by groups of two.

Exercise 3.14. Practice adding by groups. For example, take some of the additions you have already done and try doing them again breaking the second number into groups of two or three or any other convenient quantity.

3.3 Zero

Another way to think about addition is how it changes a number. For example, adding 10 to a number, n, increases n by 10. Let's look at this addition by 10 for several values of n

$$2 + 10 = 12$$
$$23 + 10 = 33$$
$$57 + 10 = 67$$
$$12 + 10 = 22$$

There is one number, 0, that does not change other numbers by its addition however. For example

$$3 + 0 = 3$$
$$0 + 2 = 2$$
$$12 + 0 = 12$$

Zero is called the *neutral element* under addition. This just means that 0 does not change a number by addition.

3.4 Order of Addition

We saw above that when we add numbers, we can add them in any order and if we add them correctly then we will get the same value. This is easy to see as indicated in figure 3.2 showing the sum $3 + 2$ using tally marks.

$$III + II = IIIII$$

There are five marks on each side of the equal sign. Three marks plus two marks on the left and five marks on the right. Three plus two is five.

Figure 3.2. The order of addition does not change the sum.

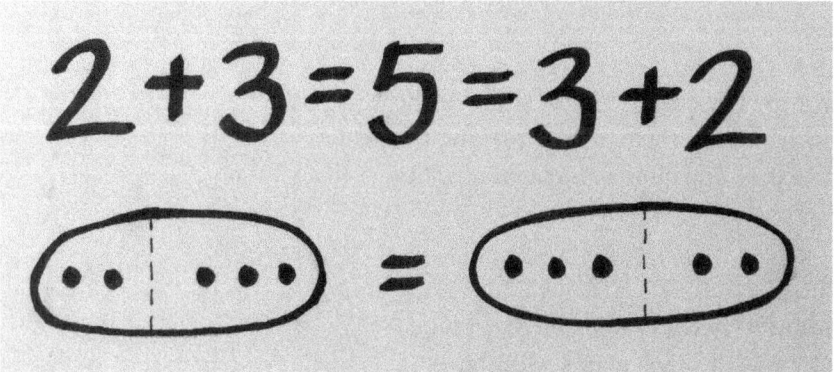

Addition is simply considering together the groups expressed on the left hand side of the equal sign. As you are just bringing the two groups together it does not matter whether you start with the group of two and add three or if you start with the group of three and add two. It is important to remember that you can change the order without changing the result of the addition.

We call this property *commutativity* and we say that addition is commutative. This means that for any numbers x and y we have

$$x + y = y + x \qquad \text{(commutativity)}$$

48 CHAPTER 3. ADDITION

For example

$$9 + 12 = 21$$
$$12 + 9 = 21$$

One way that you might use commutativity in addition as a practical matter is simply to remember that if you find it easier to add a pair of numbers one way rather than the other, you can just reorder the pair to suit your preference.

Exercise 3.15. Choose a few pairs of numbers and add each pair in both orders to check your answers and to become comfortable with this idea.

Solution:

$$18 + 9 = 27$$
$$9 + 18 = 27$$

3.4.1 Adding More Than Two Numbers

You can add up more than two numbers by keeping track of one addition at a time until you have combined all the numbers into a single count. Parentheses can be useful in showing the order of your work. For example

$$2 + 4 + 6 + 8 =$$
$$(2 + 4) + 6 + 8 = \quad \text{(parentheses show we add } 2 + 4 \text{ here)}$$
$$6 + 6 + 8 =$$
$$(6 + 6) + 8 =$$
$$12 + 8 = 20$$

Exercise 3.16.
$$3 + 5 + 7 =$$

Exercise 3.17.
$$4 + 6 + 8 =$$

Exercise 3.18.
$$7 + 3 + 9 =$$

Exercise 3.19.
$$6 + 12 + 4 =$$

Exercise 3.20.
$$14 + 2 + 8 =$$

Exercise 3.21.
$$25 + 3 + 7 =$$

Exercise 3.22.
$$18 + 5 + 9 =$$

Exercise 3.23. Choose 3 numbers and add them.

Exercise 3.24. Choose 4 numbers and add them.

Exercise 3.25. Choose 5 numbers and add them.

Figure 3.3. Showing step by step how to add five numbers.

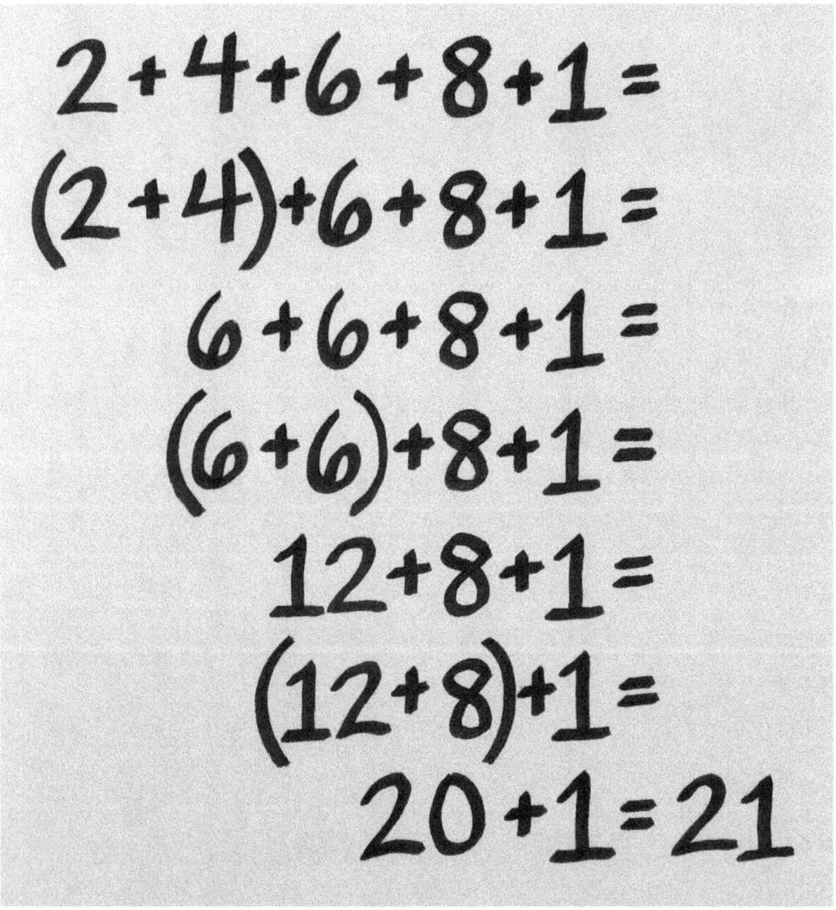

$$2+4+6+8+1=$$
$$(2+4)+6+8+1=$$
$$6+6+8+1=$$
$$(6+6)+8+1=$$
$$12+8+1=$$
$$(12+8)+1=$$
$$20+1=21$$

3.5 Counting Groups

We learned that our number system is based in counting groups of tens, the powers of ten. Each digit in a number counts a power of ten.

For example seven hundred twenty eight means seven hundreds, two tens, and eight ones

$$728 = 7 \cdot 100 + 2 \cdot 10 + 8 \cdot 1$$

The number 728 has three digits: 7, 2, and 8. We can think of this list of digits as a list of counts for each power of ten. With the number 728 we are counting hundreds, tens, and ones. So we can think about addition also as a way to help count groups and we use this to represent numbers in our base ten system.

There are a few points to pay special attention to. Let's look more closely at some numbers.

The number 10. Ten means we have ten ones. We can also think of this as one group of ten. That corresponds to how we write the number. In the number 10 the first digit counts tens and the second digit counts ones. In writing the list of digits '10' we mean *one ten and zero ones*.

The number 100. Looking at this list of three digits we see that we can think of it also in several different ways. We might think of it as 100 ones. Or we might think of it as 10 tens. Or we might think of it as 1 hundred. In fact this is what we mean when we write the number. We mean *one hundred, zero tens, and zero ones*. But it is important to appreciate that we can also think of this number as counting groups of ten. If we want to focus on tens then this number tells us that we have ten tens.

Let's now add a few numbers.

$$23 + 14 + 45$$

If we consider the *ones* in each of these three numbers we have

$$3 + 4 + 5$$

That is, we have twelve ones. Figure 3.4 shows this addition. Of course twelve ones equals one ten and two ones

$$\text{twelve ones} = 1 \cdot 10 + 2 \cdot 1$$

When we next consider how many tens we have, *we must include the one 10 from the ones that we counted*. How many tens do we have?

$$(2 + 1 + 4) + 1 = 7 + 1 = 8$$

The last ten counted comes from the ones that we counted; we had found 1 ten and 2 ones.

Figure 3.4. Adding the ones counts in the sum of 23, 14, and 45.

$$23 + 14 + 45 =$$
$$20 + 10 + 40 + 12 = 82$$

Thus our sum is 8 tens and 2 ones or 82.

Another way to write this is in column form as indicated in figure 3.5. The column form can be useful because it lets us line up the digits counting the same power of ten. But it is important to keep in mind what each digit counts. The method of column addition is included here because you might encounter it in other places. But you can also add (for example in the exercises below) using the horizontal method described above focusing on what each digit counts. Much more important than one method or another is to understand what you are counting. And if you understand the idea of counting groups in the horizontal method, then the column method will be easy.

Note that we include above the numbers in the tens column, the one ten we got from adding up the ones column.

$$3 + 4 + 5 = 7 + 5 = 12 = 10 + 2$$

We can use this method to add whole numbers of any size. Just be sure to line up the powers of ten correctly. And of course think carefully what each digit counts.

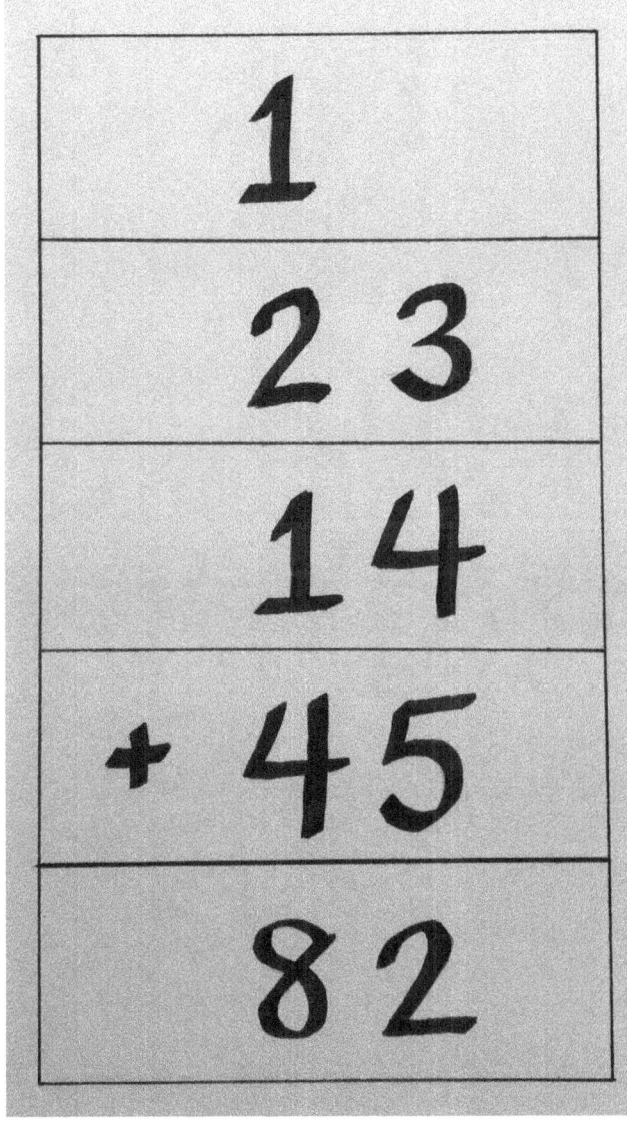

Figure 3.5. Stacking numbers to add.

54 CHAPTER 3. ADDITION

Exercise 3.26. Compute the sum

$$34,531 + 54,673$$

Solution:

See figures 3.6 and 3.7 for the details of this addition. Figure 3.7 has labels for each column.

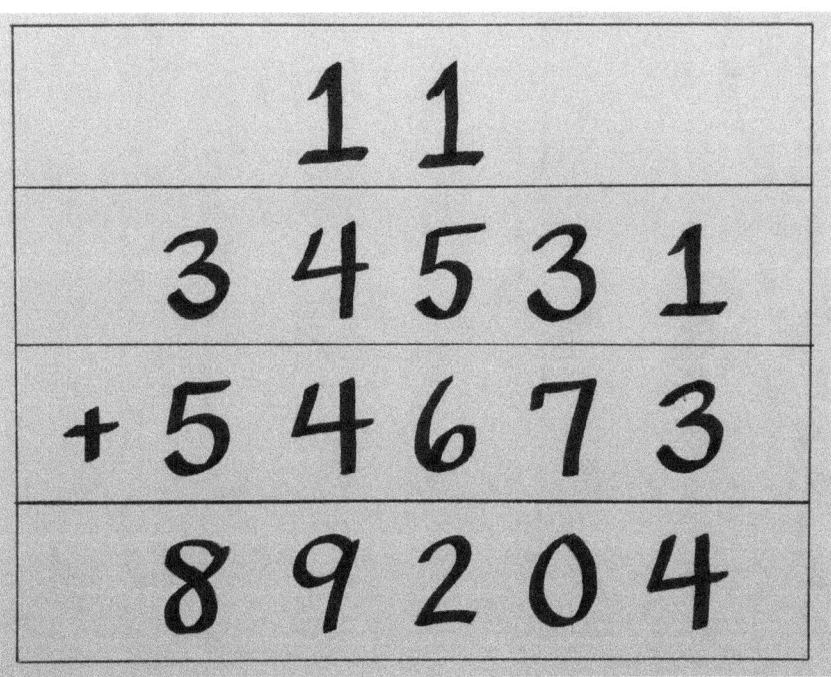

Figure 3.6. Line up the powers of ten to make adding easy.

$$34,531 + 54,673 = 89,204$$

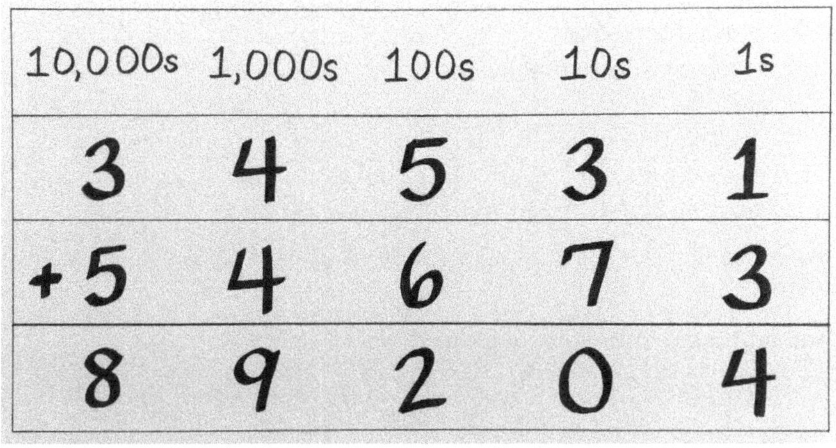

Figure 3.7. Addition with columns labelled.

10,000s	1,000s	100s	10s	1s
3	4	5	3	1
+5	4	6	7	3
8	9	2	0	4

3.5.1 *Exercises. Adding Numbers by Combining Counts of the Same Powers of Ten.*

Exercise 3.27. What is $14 + 5$?

Solution:
   14 means one ten and four ones.
   5 means five ones.
   Adding up the counts for ones we get $4 + 5$. That is 9.
   Then $10 + 9$ equals 19.

Exercise 3.28. What is $8 + 4$?

Solution:
   $8 + 2$ gives us 10.
   And $10 + 2$ equals 12.

Exercise 3.29. What is $16 + 13$?

Solution:
   One ten and one ten. The total count there is two tens. 20.
   Six ones and three ones. Total count 9 ones.
   So all together: $20 + 9$. That is 29.

Exercise 3.30. What is $17 + 18$?

Solution:
> One ten and one ten. Two tens. 20.
> Seven ones and eight ones. 15 ones. One ten and five ones.
> So together: Two tens and one ten, that is three tens. And five ones.
> 35.

Exercise 3.31. Compute $56 + 74$.

Exercise 3.32. Compute $236 + 154$.

Exercise 3.33. Compute $167 + 233$.

You can choose pairs of numbers and add them until you are perfectly comfortable adding numbers.

3.6 Number Patterns and Structures

One way to become comfortable working with numbers is to notice patterns and structures. We will consider here some patterns but you should explore others on your own. There is no single way to do this. See where your intuition takes you.

Ways to count eights

2×8. Four less than 20. 16. For example you might think $8 + 10$. And then subtract 2. For 16.

3×8. $8 + 8 + 8$. $16 + 8$. $20 + 4$. 24.

4×8. $8 + 8 + 8 + 8$. $24 + 8$. That is $3 \times 8 + 8$. And $8 + 4$ is two more than 10. That is 12. $20 + 12$. 32.

5×8. $4 \times 8 + 8$.

$40 = 5 \times 8$. Note that $8 = 2 \times 4$. So $40 = 5 \times 2 \times 4$.

$6 \times 8 = 5 \times 8 + 8$.

And so on.

Note $56 + 8 = 56 + 4 + 4 = 60 + 4$.

Up to 10×8.

Example 3.3. Counting a number, for example 8, reveals the pattern of multiples within the whole numbers.

Exercise 3.34. Choose some numbers and count up with those numbers in the same way that we counted eights above.

3.6. NUMBER PATTERNS AND STRUCTURES

3.6.1 Computing Expressions with Parentheses

This is a quick reminder that parentheses can be useful to give or to show a structure in a computation. For example in

$$5 \cdot (5 + 2 + 10)$$

the parentheses let us indicate that the additions should be computed prior to the multiplication by five.

Exercise 3.35. Compute $(3 + 4 + 6) \cdot 2$.

Exercise 3.36. Compute $(2 + 8 + 4) \cdot 4$. If necessary remember that $4 = 2 \times 2$.

Exercise 3.37. Compute $(9 + 6 + 3) \cdot 3$.

Exercise 3.38. Compute $(10 + 22 + 34) \cdot 10$.

3.6.2 Even and Odd Numbers

If you can break a whole number n into two equal whole numbers then n is said to be *even*. Otherwise n is said to be *odd*. Every whole number is thus even or odd.

If you look at numbers for a while, you will notice that every even number ends in 0, 2, 4, 6, or 8. If a whole number ends in 1, 3, 5, 7, or 9 then it is odd.

---

18 is even. You can think of it as two nines.
$9 + 9 = 18$.
But 17 is odd. There is no number k such that $k + k = 17$.
To see this consider that
$8 + 8 = 16$. But $9 + 9 = 18$. We jump over 17.

Example 3.4. The even numbers form a basic and very useful pattern (and structure) within the whole numbers. So do the odd numbers.

Even numbers are handy if you need to break a whole number into two groups of equal size.

Exercise 3.39. Choose a few two and three-digit numbers and determine whether they are even or odd. If they are even show why.

For example 512 is even because it ends in a two. We can also write 512 as the sum of two equal numbers: $512 = 256 + 256$. But 515 is odd because the ones digit, 5, is odd.

3.6.3 Powers of Ten

We saw previously (see section 1.3) how our number system counts powers of ten. It is a convenient system because if you know one through nine and understand the powers of ten then it is simple to count and track even very large numbers.

58   CHAPTER 3. ADDITION

Exercise 3.40. Write down a few numbers as lists of five or six digits. Express each number as a sum of powers of ten.

Solution:

$$452,654 = 4 \cdot 100,000 + 5 \cdot 10,000 + 2 \cdot 1,000 + 6 \cdot 100 + 5 \cdot 10 + 4 \cdot 1$$

Exercise 3.41. In the number $1,010$, discuss what are the powers of ten being counted.

Exercise 3.42. In the number $1,202$, discuss the powers of ten being counted. You can also create more exercises like this to become comfortable understanding the meaning of numbers.

3.6.4 Teens

We saw previously there is a little quirk of the English language worth noticing as we learn to use numbers. Here's how we say the teens:

thirteen for 13
fourteen for 14
fifteen for 15
sixteen for 16
seventeen for 17
eighteen for 18
nineteen for 19

In contrast, for the other decades we say the name for the decade first. For example we say

twenty-nine for 29
fifty-two for 52

Exercise 3.43.
$$9 + 2 =$$

Exercise 3.44.
$$19 + 2 =$$

Exercise 3.45.
$$29 + 2 =$$

Exercise 3.46.
$$89 + 2 =$$

Exercise 3.47.
$$59 + 2 =$$

Exercise 3.48.
$$19 + 2 =$$

Exercise 3.49.
$$2,424 + 1,000 =$$

Exercise 3.50.
$$60 + 6 =$$

Exercise 3.51.
$$60 \text{ million } + 6 \text{ million } =$$

3.7 Adding Groups of Ten

The following method might help with adding two digit numbers. It is intended only as a reference. If you have a different method that works, so much the better.

Let's say you have two numbers to add.

For example, how might we compute specifically step by step

$$54 + 79 =$$

- Identify the number with the bigger ones digit. Here it is 79 with nine ones.

- Determine what you need to add to this ones digit to reach 10. Here you need to add one because $9 + 1 = 10$.

- Express the ones digit of the other number in terms of what you found in step two. Here we write the other ones digit, four, as $4 = 1 + 3$ so that four is expressed with the one that we need to add to nine.

- Note the remaining ones. Here we have three.

- Now we see that when we add the ones digits we get one 10 and one three. Carry the one 10 over to the tens digit of one of the numbers. Let's add the new ten to the five tens of 54 so we will keep track of 6 tens there.

- Repeat the above steps to add the tens: six tens plus seven tens: first we note that $7 + 3 = 10$ so write $6 = 3 + 3$. We then add $3 + 3 + 7 = 3 + (3 + 7) = 3 + 10 = 13$. That is thirteen tens so we have 130.

- Finally add the tens digit to the ones digit: $130 + 3 = 133$.

Another way to do this is to start with the tens:

- Add 5 and 7 for the tens digit. Seven is three short of 10 so write five in terms of three: $5 + 7 = 2 + 3 + 7 = 2 + (3 + 7) = 2 + 10 = 12$. So we have 12 tens. 12 tens is 120.
- Add the four and the nine of the ones count. $4 + 9 = 3 + 1 + 9 = 3 + 10 = 13$. So we have $120 + 13$.
- Repeat to get $120 + 13 = 133$.

The most important idea: think what each digit counts.

3.7.1 Additional Suggestions

Practice adding pairs that add to ten so you recognize them easily. For example, if you want to compute the sum

$$2 + 8 + 6 + 4 + 3 + 7 + 1$$

recognize that $2 + 8$, $6 + 4$, and $3 + 7$ each equals 10 so that the sum total is 31.

$$5 + 5 = 10$$
$$4 + 6 = 10$$
$$3 + 7 = 10$$
$$2 + 8 = 10$$
$$1 + 9 = 10$$

When you add numbers ending with nine simply take one from the number you are adding so that your number ending with nine turns into a *round number* (if that is easier). A round number is a number that ends in zero. 30, 160, 1,000 are round numbers. Round numbers are often easy to work with.

$59 + 3 = 59 + 1 + 2 = (59 + 1) + 2 = 60 + 2 = 62$ so when you have an addition like $59 + 3$ think of it as $60 + 2 = 62$.

3.7.2 Advice for More Practice

At this point fluency with addition is a matter of practice. Here is guidance for a sequence of exercises.

First learn well the pairs of numbers adding to ten.

Practice adding numbers changing the order of addition. Add numbers from left to right and from right to left.

It is important to develop the habit of recognizing different ways to express the same sum and of using prior results to find quick answers for new questions.

Build up your understanding of how to write numbers by adding pairs of two digit numbers and then pairs of three digit numbers first where there is no need to carry over in the addition. Here the sum of corresponding digits is always less than 10. For example

$$264 + 325 = 589$$

As you become more familiar with the simple additions, start to work with bigger numbers. Practice two digit additions that require carrying a ten over from the ones to the tens digits. Start this by adding numbers to two digit numbers ending in nine. For example $3 + 89 = 92$.

Next start with one three digit number and add 1, 10, 100, 101, and 11 and so on.

Another way to confirm that you understand the meaning of digits in a number is to add numbers that combine only different powers of ten. For example $700 + 40 = 740$ or $1,000 + 101 = 1,101$. These exercises focus on distinguishing the powers of ten.

Answer the following questions and create additional questions similar to these.

Exercise 3.52.
$$5 + 5 =$$

Exercise 3.53.
$$8 + 2 =$$

Exercise 3.54.
$$50 + 50 =$$

Exercise 3.55.
$$500 + 500 =$$

Exercise 3.56.
$$5,000 + 5,000 =$$

Exercise 3.57.
$$50,000 + 50,000 =$$

Exercise 3.58.
$$500,000 + 500,000 =$$

Exercise 3.59. Add pairs of numbers that require carrying up powers of ten repeatedly. For example

$$777 + 223 = 1,000$$

This addition requires carrying up a ten from the ones and a hundred from the tens.

4 Multiplication

4.1 Multiplication as Repeated Addition

As a factual matter multiplication is *repeated addition*. It can be thought of as just another way to express addition. We can also think about multiplication in different ways as we will see later, but repeated addition is a sensible place to start. And, in a sense, multiplication is a notational convenience. It is a more concise expression of repeated additions.

We commonly use either of two symbols to indicate multiplication '$\times$' and '$\cdot$'. Both symbols mean the same idea. The result of a multiplication is often called a *product*.

We write

$$3 \times 4 = 12 \quad \text{or}$$
$$3 \cdot 4 = 12$$

and we say *three times four equals twelve*. Literally *three times four* means that we count four three times. One four is four, two fours are eight, three fours are twelve. We see in these countings that, in the first instance, multiplication corresponds to repeated addition.

$$3 \cdot 4 = 4 + 4 + 4$$
$$= (4 + 4) + 4$$
$$= 8 + 4$$
$$= 12$$

Here we see *addition* of four repeated twice. We add $4 + 4$ and then we add four again ($8 + 4$) so that we have added three fours together.

The product of three times four is twelve.

$$3 \cdot 4 = 12$$

As we will see, the multiplication notation is especially convenient with big numbers. For example

$$4 \times 100$$

We could think of this as one hundred fours or four hundreds.

$$4 + 4 + 4 + 4 + \cdots + 4 \quad \text{(one hundred fours)}$$

This is the same as four one hundreds

$$100 + 100 + 100 + 100 = 400$$

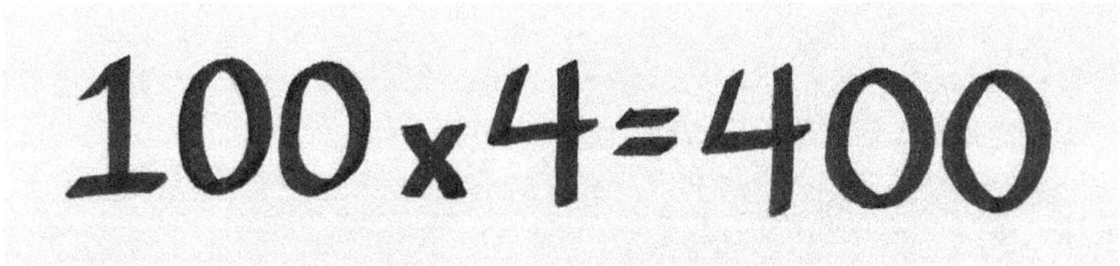

Figure 4.1. Multiplication as a convenient notation, especially for adding large numbers.

$$100 \times 4 = 400$$

4.2 Multiplication as Counting Groups

We can also think of multiplication as *counting groups* of a number. When we multiply three by seven we are counting three groups of seven (or seven groups of three) and we might say we are counting three sevens.

In this perspective, think of the quantity (seven) as a unit and then count up how many units we have. A *unit* is like a single building block that we designate and then use to count. We can choose units that are convenient. If it is convenient to count in groups of seven we can choose a unit with seven ones (or seven other items). If our unit is seven ones then every time we count one unit, we are counting seven ones.

$$3 \cdot 7 = (7) + (7) + (7)$$
$$= 14 + (7)$$
$$= 21$$

The parentheses are used here merely to mark the units to count. We are counting sevens. Each seven is one of our units to count.

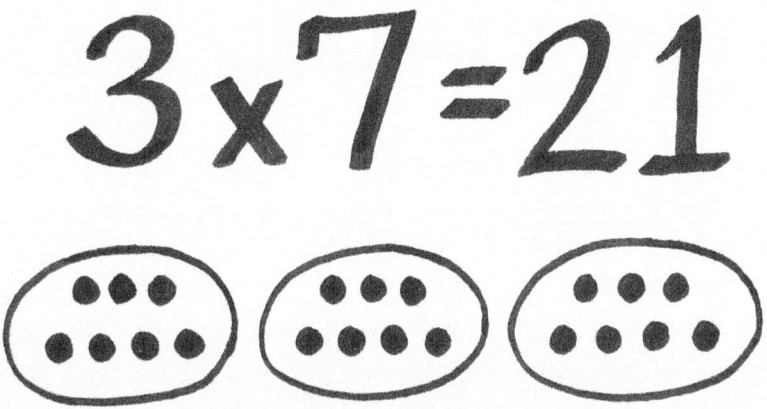

Figure 4.2. Counting 21 with sevens. Our unit is seven. Three sevens are 21.

Let's go back to our three times four example step by step. We will use parentheses this time to show each arithmetic step:

$$3 \cdot 4 =$$
$$4 + 4 + 4 =$$
$$(4 + 4) + 4 =$$
$$8 + 4 = 12$$

66 CHAPTER 4. MULTIPLICATION

Similarly

$$4 \cdot 2 =$$
$$2+2+2+2 =$$
$$(2+2)+2+2 =$$
$$(4+2)+2 =$$
$$6+2 = 8$$

And

$$3 \cdot 5 =$$
$$5+5+5 =$$
$$(5+5)+5 =$$
$$10+5 = 15$$

Let's review the meaning of the equal sign ('=') in these statements. The equal sign indicates that the expression on the left hand side has the same value as the expression on the right hand side. So notice that each equality shows us two different ways to express the same value. Indeed there are often many ways to express a given value. We use this fact in simplifying computations.

Exercise 4.1. Compute $12 \cdot 8$

Solution:

$$\begin{aligned}12 \cdot 8 &= (10+2) \cdot 8 \\ &= 10 \cdot 8 + 2 \cdot 8 \\ &= 80 + 16 \\ &= 96\end{aligned}$$

There are many ways to express a given relationship. Multiplication illustrates this. And keep in mind that there are likewise many ways to think about these kinds of equalities.

Exercise 4.2. Compute $14 \cdot 7$

Solution: Here is one way to approach this. Think of this multiplication as counting groups of seven. We want to count 14 groups of seven. Let's think of 14 in easy pieces: 10 plus 4. Fourteen groups of seven is equal to ten groups of seven and four more groups of seven.

$$14 \cdot 7 = (10+4) \cdot 7 = (10 \cdot 7) + (4 \cdot 7) = 70 + 28 = 98$$

Exercise 4.3. Compute $17 \cdot 3$

Exercise 4.4. Compute $9 \cdot 11$

Exercise 4.5. Compute $15 \cdot 9$

Exercise 4.6. Compute $3 \cdot 17$

Exercise 4.7. Compute $13 \cdot 13$

Exercise 4.8. Compute $19 \cdot 6$

Exercise 4.9. Compute $14 \cdot 8$

Exercise 4.10. Compute $12 \cdot 9$

Exercise 4.11. Compute $16 \cdot 6$

Exercise 4.12. Compute $29 \cdot 3$

Exercise 4.13. Compute $18 \cdot 7$

Exercise 4.14. Compute $50 \cdot 28$

Exercise 4.15. Compute $6 \cdot 9$

Exercise 4.16. Compute $13 \cdot 15$

Exercise 4.17. Compute $13 \cdot 23$

Exercise 4.18. Compute $17 \cdot 6$

Exercise 4.19. Devise and compute additional multiplications; as many as necessary to become comfortable with these kinds of multiplications.

4.3 *Powers of Ten and Other Group Multiplications*

The way we write numbers, based in the powers of ten, also illustrates a specific use of multiplication. The digits of a number tell us the quantity of each of power of ten required to form the number.

For example the number three thousand and two is written $3,002$. The digits in this number tell us that it represents three thousands plus zero hundreds, plus zero tens, plus two. That is three thousands and two ones. We write

$$3,002 = 3 \cdot 1,000 + 0 \cdot 100 + 0 \cdot 10 + 2 \cdot 1$$

The *comma* (',') in $3,002$ is just a convenience to separate three digits at a time so the list of digits is easier to read. Marking out every three digits (from the right) with a comma—like this $7,653,005$—makes a number easier to read than an unmarked list of digits (7653005).

With multiplication we extend our counting of groups from powers of ten to quantities of any number.

$$3 \cdot 100 + 2 \cdot 10 = 320$$

This counts three hundreds and two tens.

$$3 \cdot 6 + 2 \cdot 4 = 18 + 8$$
$$= 26$$

Similarly twenty-six can be thought of as comprising three sixes and two fours.

It is often useful to think of the same number in different forms and expressions.

Consider the number 20.
  We can think of 20 as two tens.
  We can also think of 20 as four fives.

Example 4.1. There are different ways to think of the number 20. For example as two tens or as four fives.

Exercise 4.20. Choose a few numbers and express each number in at least two different ways.

4.4 Multiplication Patterns and Structures

As we learn multiplication, there are many patterns to notice based in the definition of multiplication as repeated addition. Let's consider a few of these patterns.

4.4.1 Tens

$$1 \cdot 10 = 10$$
$$2 \cdot 10 = 20$$
$$3 \cdot 10 = 30$$
$$4 \cdot 10 = 40$$
$$5 \cdot 10 = 50$$
$$6 \cdot 10 = 60$$
$$7 \cdot 10 = 70$$
$$8 \cdot 10 = 80$$
$$9 \cdot 10 = 90$$
$$10 \cdot 10 = 100$$
$$12 \cdot 10 = 120$$

You can see that we are simply counting up tens as we multiply 10 by bigger numbers. As a formal matter multiplying any number n by ten merely requires sticking a zero on the right of n. This makes sense because n times ten means counting n tens. The digits to the left of one zero count tens.

---

10 counts one ten.

 70 tells us we have seven tens.

 110 tells us we have eleven tens.

 1,100 tells us we have 110 tens.

Example 4.2. Notice the patterns in counting with tens.

More generally you can think of big round numbers as counting by groups of other powers of ten as well.

 1,100 counts eleven hundreds.

 150,000 counts 150 thousands.

 1,600,000 counts 1,600 thousands.

Exercise 4.21. Play around with counting big numbers (and big powers of ten) in terms of different powers of ten, say 100, 1,000, 10,000.

Exercise 4.22. For example 1,000,000 is one thousand hundreds or 100,000 tens or ten 100,000s or 10,000 hundreds. Look for a pattern as you consider big powers of ten as sums of smaller powers of ten.

Exercise 4.23. Practice multiplications, deliberately thinking about how to compute the given multiplication. For example, 6×7. You might think, *let's count three sevens. I can reuse the computation. Three sevens and three sevens equal six sevens.*

4.4.2 Fives

$$1 \cdot 5 = 5$$
$$2 \cdot 5 = 10$$
$$3 \cdot 5 = 15$$
$$4 \cdot 5 = 20$$
$$5 \cdot 5 = 25$$
$$6 \cdot 5 = 30$$
$$7 \cdot 5 = 35$$
$$8 \cdot 5 = 40$$
$$9 \cdot 5 = 45$$

Note patterns in the multiples of five.

4.4.3 Elevens

$$1 \cdot 11 = 11$$
$$2 \cdot 11 = 22$$
$$3 \cdot 11 = 33$$
$$4 \cdot 11 = 44$$
$$5 \cdot 11 = 55$$
$$6 \cdot 11 = 66$$
$$7 \cdot 11 = 77$$
$$8 \cdot 11 = 88$$
$$9 \cdot 11 = 99$$
$$10 \cdot 11 = 110$$
$$12 \cdot 11 = 121$$

4.4.4 Twos

$$2 \cdot 1 = 2$$
$$2 \cdot 2 = 4$$
$$2 \cdot 3 = 6$$
$$2 \cdot 4 = 8$$
$$2 \cdot 5 = 10$$
$$2 \cdot 6 = 12$$
$$2 \cdot 7 = 14$$

Exercise 4.24. Write ten multiples of three and ten multiples of five.

Exercise 4.25. Note patterns in the multiples of eleven.

4.4.5 Looking At The Powers of Ten Again

The *powers of* 10 are just particular, though important, multiples of 10. The powers of 10 are produced by multiplying 1 or a power of ten by ten or a power of ten. It is worthwhile to spend some time with them and to become comfortable thinking about them in different ways.

Figure 4.3 shows the first nine powers of 10. Notice how these are just multiples of 10.

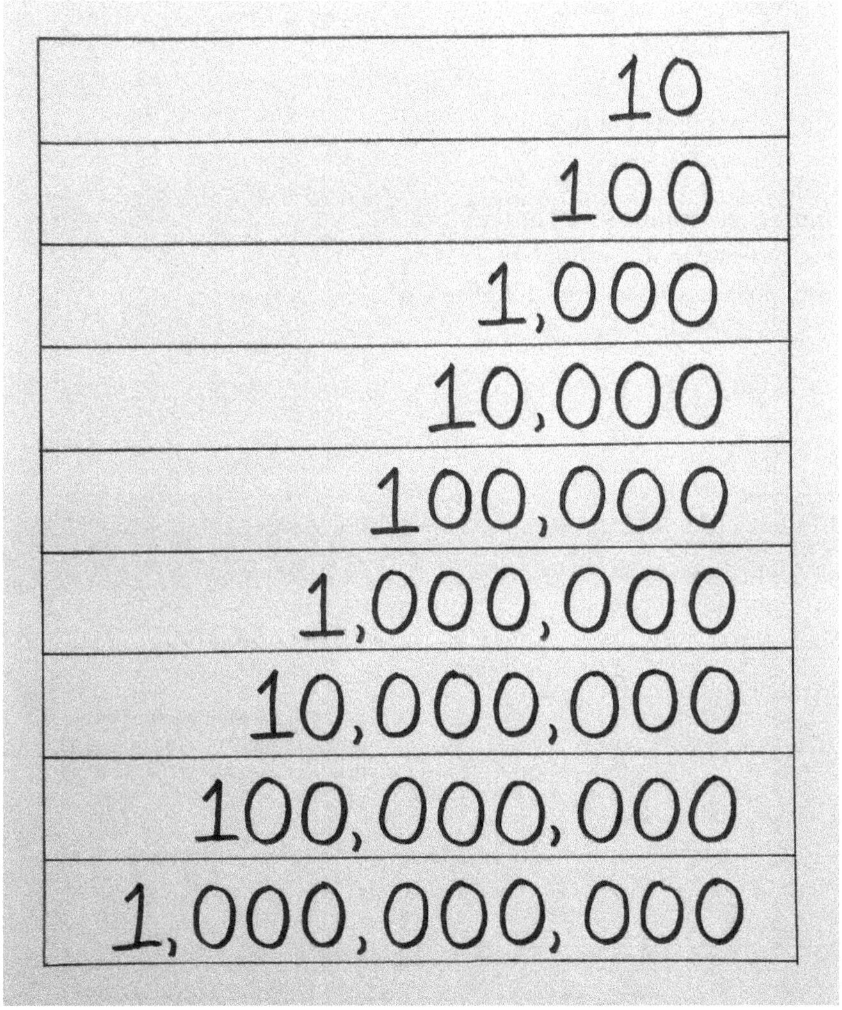

Figure 4.3. The first 9 powers of 10.

10
100
1,000
10,000
100,000
1,000,000
10,000,000
100,000,000
1,000,000,000

4.4.6 A Note on Thinking of New Expressions

Let's take for example the number 35. Let's say that we want to express this number as a multiplication. First note that it ends in five. If we also know that 35 is a multiple of 7, then $5 \cdot 7$ is a good guess because 35 must also be a multiple of five. Another example:

If you have computed $3 \cdot 6 = 18$ and you wish to solve $6 \cdot 6$, just *reuse* your prior work, noticing that $6 \cdot 6$ is simply $(3 \cdot 6) + (3 \cdot 6)$.

Example 4.3. We can use smaller multiplications to compute bigger multiplications.

Exercise 4.26. Write out multiples of other numbers not listed above and discuss their patterns.

Exercise 4.27. Choose a few two or three digit numbers, n. Find multiplication expressions for these numbers or indicate that you think there is no multiplication possible to express your number other than the multiplication $1 \cdot n$.

4.4.7 Multiplication By One

For any number n

$$1 \cdot n = n \quad \text{and}$$
$$n \cdot 1 = n$$

$$1 \cdot 12 = 12$$
$$100 \cdot 1 = 100$$
$$4 \cdot 1 = 4$$

Example 4.4. Multiplication by one can be thought of as an operation that does not change the other factor.

Another way to think about this is that multiplying a number n by 1 does not change n. This follows from the definition of multiplication. $1 \cdot n$ simply means count n once. In particular you don't add anything to n. Because multiplication by one does not change n we say that the number one is the *identity* element for multiplication.

Recall that the identity for *addition* is 0. If you add 0 to any number n, the result is always still n.

4.4.8 Note on Using Letters in Arithmetic Expressions

It is often convenient to use letters, like n or x, as a name for numbers when we talk about numbers generally rather than about specific numbers. For example here we are discussing generally what happens when we multiply a number by one. So it is convenient to give a name to the number that we are discussing. In this way the property we are thinking about should be understood to apply to all the numbers that we are considering. This is a common and useful practice.

4.4.9 Multiplication By Zero

To multiply a number n by zero means that we take that number n no times. What do we have then? Nothing. Multiplication by zero is always zero.

For any number n

$$0 \cdot n = 0 \quad \text{and}$$
$$n \cdot 0 = 0$$

$$0 \cdot 10 = 0$$
$$10 \cdot 0 = 0$$
$$25 \cdot 0 = 0$$
$$0 \cdot 1 = 0$$
$$1 \cdot 0 = 0$$

Example 4.5. Counting 0 anything is always 0. So multiplication by zero always yields 0.

4.4.10 Zero for Multiplication and for Addition

To be clear about the different effects of zero for multiplication and for addition let's look these together now. When we *add zero* to a number we do not change the number: $n + 0 = n$. But when we multiply 0 with a number we always get 0. $n \cdot 0 = 0$. The case of addition is easy to think about. We have a number and we are adding nothing. There is no change. The number stays the same. For the case of multiplication, which we can think of as a counting of a group, multiplication is in effect counting no group. If we think of multiplication as a way to change a number, then multiplication by 0 neutralizes all numbers. Multiplication by zero turns numbers into 0. What is the neutral element in this sense for addition? There isn't a single neutral element but we can neutralize numbers in addition by using the negative of the number. $k + -k = 0$. Keep in mind that if k is less than zero then $-k$ is positive. To review negative numbers, see section 2.2. We will also discuss negative numbers in detail in chapter 6.

4.4.11 Multiplication Tables

One way to learn multiplication is to write out a multiplication table. This is not necessary, but it might be helpful. You can learn multiplication by playing around with numbers directly. If you find a multiplication table useful then make one. The memorization will not hurt but also practice thinking through the multiplications based in the meaning of numbers and multiplication.

If you do use a table, focus on patterns rather than rote memorization. The basic idea of a multiplication table is that each table box below the top row and to the right of the left column expresses the product of the box's row number with its column number.

With or without tables, become familiar with multiplications at least up to $20 \cdot 20$.

4.5 The Property of Commutativity $n \cdot m = m \cdot n$

For any numbers n and m we have

$$n \cdot m = m \cdot n$$

This property of multiplication is called *commutativity*.

Counting m groups of n produces the same result as counting n groups of m. So for example if you know the answer to $n \cdot m$ then you also know the answer to $m \cdot n$. Because of this we say that multiplication is *commutative*. The name is not so important but you should certainly understand the meaning.

$$11 \cdot 5 = 55 \quad \text{and}$$
$$5 \cdot 11 = 55$$

Example 4.6. In multiplying two numbers, the order does not change the result.

This is a pretty neat property. It says for example

$$4 + 4 + 4 + 4 + 4 = 5 + 5 + 5 + 5$$

Indeed

$$5 + 5 + 5 + 5 = (4+1) + (4+1) + (4+1) + (4+1)$$
$$= 4 + 4 + 4 + 4 + (1 + 1 + 1 + 1)$$
$$= 4 + 4 + 4 + 4 + 4$$

You can convince yourself, as indicated above, that this makes sense by counting groups and expressing the bigger number in terms of the smaller number. In the example above we could write $(4 + 1)$ instead of each 5. Then we could group together all the ones. We would then have another group of fours so that we would have five fours.

4.6 The Property of Associativity $(n \cdot m) \cdot p = (m \cdot p) \cdot n$

The order of the multiplications in a series does not change the answer. This is the called the *associative* property of multiplication.

$$3 \cdot 5 \cdot 100 = 3 \cdot 500$$
$$3 \cdot 5 \cdot 100 = 5 \cdot 300$$
$$3 \cdot 5 \cdot 100 = 15 \cdot 100$$

Convince yourself that this makes sense. Think of counting groups as we did with the commutative property.

Exercise 4.28. Compute $8 \cdot 5$. Then use that result to compute $8 \cdot 50$. What about $8 \cdot 500$? Choose other numbers and create more calculations like these three.

4.7 Building Up Answers With Easier Questions: Distributivity

When we compute a multiplication of big numbers, it is often convenient to break the numbers into a sum of smaller numbers. Then we can compute the multiplication in smaller parts. This practice also helps to learn multiplication and addition. For example it is often helpful to break up a number into a sum including 10 because 10 is easy to work with.

If we think of the 17 as contained in a bag we can imagine the 17 in two pieces. One 10 and one 7. If we take 8 of these bags then it is easy to see that we have taken eight tens and also eight sevens.

Concerning the notation in the expression $10 \cdot 8 + 7 \cdot 8$ we do not need to use parentheses because multiplication takes priority over addition so it is understood that we multiply in the expression before we add. Because this precedence corresponds to our intention in this expression parentheses are not needed.

Multiply 17×8.

$$\begin{aligned} 17 \cdot 8 &= (10+7) \cdot 8 \\ &= 10 \cdot 8 + 7 \cdot 8 \\ &= 80 + 56 \\ &= 136 \end{aligned}$$

Example 4.7. Distributivity is just a technical term for the idea that we can count multiples in different ways without changing the result. For example we can count 17 eights by counting ten eights and seven eights.

Here is another example featuring breaking a number into a sum and using parentheses to group the terms of the sum.

$$15 \times 7 = (11+4) \times 7$$

Example 4.8. To count 15 sevens we can count first 11 sevens and then four sevens.

We can break big numbers down into smaller numbers to simplify multiplications.

For example if we know

$$6 \cdot 6 = 36$$

then we can easily determine 6×7.

We want to count seven sixes. Seven sixes are six sixes plus one more six.

$$\begin{aligned} 6 \cdot 7 &= 6 \cdot 6 + 6 \\ &= 36 + 6 \\ &= 42 \end{aligned}$$

More generally we can write

$$n \cdot k = (n-1) \cdot k + k \quad \text{(for whole numbers } n \text{ and } k\text{)}$$

We can break down numbers in other ways too.

$$12 \cdot 6 = (10+2) \cdot 6$$

Here we write the number twelve two ways. First just as 12 and then as $(10+2)$. We use parentheses to indicate that the $10+2$ are one group and that we must do the addition *before* the multiplication. This is important because $(10+2) \cdot 6$ is not the same as $10 + 2 \cdot 6$.

$$10 + 2 \cdot 6 = 10 + 12 = 22$$

That is not what we are computing with $12 \cdot 6$ or six twelves. Rather, six twelves means we have six tens and six twos. That is

$$12 \cdot 6 = (10 + 2) \cdot 6 = 10 \cdot 6 + 2 \cdot 6$$

This last step illustrates the *distributive* property. The multiplication by six distributes to the ten and the two. Now we have two simpler multiplications and one addition to find the answer to $12 \cdot 6$.

$$12 \cdot 6 = 60 + 12 = 72$$

Thus in computing a multiplication with numbers bigger than you are used to, you can break down those numbers into smaller numbers that are easier for you to work with. This is very useful.

$$\begin{aligned} 7 \cdot 25 &= 7 \cdot (20 + 5) \\ &= 7 \cdot 20 + 7 \cdot 5 \\ &= 140 + 35 \\ &= 175 \end{aligned}$$

Example 4.9. Using smaller multiplications to compute bigger multiplications.

4.7.1 Practice Tip

Use round numbers and tens in multiplications. For example

$$\begin{aligned} 4 \cdot 35 &= 4 \cdot (30 + 5) \\ &= 4 \cdot 30 + 4 \cdot 5 \\ &= 120 + 20 \\ &= 140 \end{aligned}$$

You can get the same result a different way:

$$\begin{aligned} 4 \cdot 35 &= 4 \cdot (32 + 3) \\ &= 4 \cdot 32 + 4 \cdot 3 \\ &= 128 + 12 \\ &= 140 \end{aligned}$$

78   CHAPTER 4. MULTIPLICATION

Notice here again how we use parentheses to indicate what calculation comes first. In $4 \cdot (32 + 3)$ the parentheses tell us that we must add $32 + 3$ before multiplying.

It is useful to be comfortable expressing numbers as different forms of multiplications.

Exercise 4.29. Multiply $56 \cdot 3$

Solution:

First we break 56 into smaller numbers easier to multiply.

$$56 \cdot 3 = (50 + 6) \cdot 3$$

Then we count three fifties and three sixes. That is we distribute the three to multiply the fifty and also the six.

$$\begin{aligned}(50 + 6) \cdot 3 &= 50 \cdot 3 + 6 \cdot 3 \\ &= 150 + 18 \\ &= 168\end{aligned}$$

Exercise 4.30. Multiply $21 \cdot 10$

Solution:

$$21 \cdot 10 = (20 + 1) \cdot 10$$

This means you are counting 20 tens and one more ten; 20 tens are 200; one ten is ten; 210.

Exercise 4.31. Multiply $6 \cdot 14$

Solution:

$$\begin{aligned}6 \cdot 14 &= 6 \cdot (10 + 4) \\ &= 6 \cdot 10 + 6 \cdot 4 \\ &= 60 + 24 \\ &= 84\end{aligned}$$

Exercise 4.32. Multiply $42 \cdot 3$

Solution: Say you have a box with 42 in two pieces: a piece of 40 and a piece of 2. If you have three boxes you have three forties and three twos. You have $120 + 6$. So the answer is 126.

Exercise 4.33. Multiply $57 \cdot 5$

Exercise 4.34. Multiply $57 \cdot 15$

Exercise 4.35. Multiply $57 \cdot 20$

Exercise 4.36. Multiply $57 \cdot 19$

Exercise 4.37. Multiply $3 \cdot 27$

Exercise 4.38. Multiply $3 \cdot 30$

Exercise 4.39. Multiply $13 \cdot 27$

Exercise 4.40. Multiply $13 \cdot 270$

Exercise 4.41. Multiply $8 \cdot 62$

Exercise 4.42. Multiply $28 \cdot 62$

Exercise 4.43. Multiply $12 \cdot 62$

Exercise 4.44. Multiply $16 \cdot 62$

Exercise 4.45. Choose a few simple numbers and figure out how to produce them using multiplication.

Exercise 4.46. Choose a few numbers and multiply them together.

Exercise 4.47. How many twelves are in 72? How many pairs of twelves are in 72?

Exercise 4.48. How many twelves are in 36?

Exercise 4.49. Knowing that $3 \times 18 = 54$, determine 6×18.

Exercise 4.50. Determine $18 \times 1,000$

Exercise 4.51. Determine $18 \times 100,000$

Exercise 4.52. 31×10

Exercise 4.53. 12×7

Exercise 4.54. 13×8

Exercise 4.55. Multiply $15,000 \times 4$

Exercise 4.56. Multiply $60 \times 1,000$

Exercise 4.57. Compute $15,000 \cdot 11$

Exercise 4.58. $13,000 \cdot 12$

Exercise 4.59. $180,000 \cdot 6$

Exercise 4.60. $27 \cdot 16$

Exercise 4.61. $56 \cdot 4$

CHAPTER 4. MULTIPLICATION

Exercise 4.62. $28 \cdot 12$

Exercise 4.63. $89 \cdot 12$

Exercise 4.64. $4 \cdot 890$

Exercise 4.65. $26 \cdot 7$

Exercise 4.66. $26 \cdot 216$

Exercise 4.67. $14 \cdot 6$

Exercise 4.68. $14 \cdot 16$

Exercise 4.69. $32 \cdot 16$

Exercise 4.70. Pick a few two digit numbers. Then set out some multiplications computing each number. For example, the number 56, what are different countings we can describe resulting in 56.

Exercise 4.71. $13 \cdot 17$

When we express a number as a multiplication we are expressing the number in terms of *factors*.

In the multiplication $2 \cdot 3 = 6$, two and three are factors of six. To *factorize* a number means to express that number as a product of factors.

Example 4.10. Two and three are factors of six.

Let's find all the factors of 12.

$$3 \cdot 4 = 12$$
$$2 \cdot 6 = 12$$

2, 3, 4, and 6 are factors of 12.

Example 4.11. Find the factors of 12.

Exercise 4.72. Factorize the number 15.

Exercise 4.73. Factorize 6.

Exercise 4.74. Factorize the number 21.

Exercise 4.75. Factorize the number 36.

Exercise 4.76. Factorize the number 60.

Exercise 4.77. Practice multiplications in sets of questions where the goal is to use the answer of the first computation to help compute the answer to the next question. The idea here is to practice the understanding that we can build up answers from answers to easier questions. This prepares the subsequent practice of breaking a bigger problem into smaller parts that are easier to solve. For example. What is 3×5? Then use the answer to this question to compute 7×5. Practice this quickly sometimes. You can break the multiplications down into their corresponding additions and group the additions if that is helpful to see what is going on. Try to find clever groupings. That is, smaller multiplications that are easy for you. Then when you combine the easier multiplications you can build up the answer to the more difficult computation.

Exercise 4.78. Multiply 6×12. And then compute a few multiples of the result.

Exercise 4.79. Multiply 12×12 by breaking the numbers into sums.

Exercise 4.80. Multiply 15×15 by breaking the numbers into sums.

Exercise 4.81. Multiply 18×18 by breaking the numbers into sums.

Exercise 4.82. Multiply 18×12 by breaking the numbers into sums.

Exercise 4.83. Multiply 12×15 by breaking the numbers into sums.

Exercise 4.84. Multiply 15×18 by breaking the numbers into sums.

Exercise 4.85. If you know that $4 \times 12 = 48$, then how many twelves are in $48 + 48$ (96)?

Exercise 4.86. How many twelves do you have in 72?

Exercise 4.87. How many twelves do you have in 144 ($72 + 72$)?

Solution: To determine 12×12. This means twelve twelves. This is six twelves plus six twelves. We know that $6 \times 12 = 72$. So $12 \times 12 = 144$. You could also say $12 \times 12 = 3 \cdot 4 \times 12$. The important idea: to see how we are counting twelves and how we can build up multiplications.

Exercise 4.88. Multiply 13×6 using what we have already worked out.

Exercise 4.89. Multiply 13×12 using what we know of 6×13.

Exercise 4.90. Write out other ways to express 13×12 or 12×13.

4.8 Prime Factorization

We saw how to factorize the number 6. What about the number 7? It is easy to verify that there is not a pair of whole numbers whose product equals 7. The best we can do is the multiplication $1 \times 7 = 7$. When this is the case for a number n like it is here for the number 7, we say that n is a prime number.

In a prime factorization we factorize all the factors into primes only.

To find the prime factorization of a number, keep factorizing until all factors are prime. The prime factorization of 60 is

$$60 = 2 \cdot 2 \cdot 3 \cdot 5$$

Step by step (for example):

$$\begin{aligned} 60 &= 6 \cdot 10 \\ &= 2 \cdot 3 \cdot 2 \cdot 5 \\ &= 2 \cdot 2 \cdot 3 \cdot 5 \end{aligned}$$

Example 4.12. Find the prime factors for 60.

Exercise 4.91. Here are the first four prime numbers: 2, 3, 5, 7. What is the fifth prime number? And the sixth?

Exercise 4.92. What is the smallest prime number bigger than 20?

Exercise 4.93. What is the smallest prime number bigger than 30?

Exercise 4.94. How many numbers do you have to check to confirm that 17 is prime?

Solution: We can see that 2 is not a factor because 17 is odd. We can check 3. $3 \cdot 5 = 15$ whereas $3 \cdot 6 = 18$. So 3 is not a factor. $4 = 2 \cdot 2$ so it cannot be a factor because 17 is odd. 5 cannot be a factor because multiples of 5 end in 5 or 0. Multiples of 6 are even. $7 \cdot 2 = 14$ and $7 \cdot 3 = 21$ so 7 is not a factor. Multiples of 8 are even. $9 \cdot 2$ is bigger than 17. So after checking 2, 3, 4, 5, 6, 7, 8 and 9 we confirm that 17 must be prime.

4.8. PRIME FACTORIZATION

Exercise 4.95. Choose any whole number and then think of the different ways to express this number as a multiplication of two numbers. For example check that the following multiplications express 60.

Solution: Pairs of factors for 60:

$$6 \cdot 10$$
$$10 \cdot 6$$
$$30 \cdot 2$$
$$2 \cdot 30$$
$$20 \cdot 3$$
$$3 \cdot 20$$
$$12 \cdot 5$$
$$5 \cdot 12$$
$$15 \cdot 4$$
$$4 \cdot 15$$

Exercise 4.96. Find the prime factorization for 56.

Exercise 4.97. Find the prime factorization for 64.

Exercise 4.98. Find the prime factorization for 39.

Exercise 4.99. Find the prime factorization for 18.

Exercise 4.100. Find the prime factorization for 37.

Exercise 4.101. Find the prime factorization for 27.

Exercise 4.102. Find the prime factorization for 100.

Exercise 4.103. Find the prime factorization for $1,000$.

Exercise 4.104. Find the prime factorization for 23.

Exercise 4.105. Re-write the following three-factor factorization as two-factor factorizations. $3 \cdot 5 \cdot 10$.

Solution:

$$\begin{aligned} 3 \cdot 5 \cdot 10 &= 15 \cdot 10 \text{ or} \\ &= 30 \cdot 5 \text{ or} \\ &= 3 \cdot 50 \end{aligned}$$

Exercise 4.106. Re-write the following three-factor factorization as two-factor factorizations. $3 \cdot 4 \cdot 10$.

Exercise 4.107. Re-write the following three-factor factorization as two-factor factorizations. $4 \cdot 5 \cdot 6$.

Exercise 4.108. Re-write the following three-factor factorization as two-factor factorizations. $2 \cdot 4 \cdot 6$.

Exercise 4.109. Re-write the following three-factor factorization as two-factor factorizations. $9 \cdot 8 \cdot 7$.

Exercise 4.110. Given $4 \cdot 13$, express this product with different factors.

Exercise 4.111. Given $5 \cdot 12$, express this product with different factors.

Exercise 4.112. Given $3 \cdot 16$, express this product with different factors.

Exercise 4.113. Given $8 \cdot 7$, express this product with different factors.

5 Division

5.1 From Multiplication To Division

We will approach division initially from what we know about multiplication. Let's start with a few simple multiples of ten.

$$1 \cdot 10 = 10$$
$$2 \cdot 10 = 20$$
$$3 \cdot 10 = 30$$
$$4 \cdot 10 = 40$$
$$5 \cdot 10 = 50$$
$$6 \cdot 10 = 60$$
$$11 \cdot 10 = 110$$

We know that we can think of these multiplications as repeated additions. For example,

$$4 \cdot 10$$

counts up four tens.

$$10 + 10 + 10 + 10 = 40$$

We also know that the expression $4 \cdot 10$ equals $10 \cdot 4$ so that we could have counted up ten fours.

These multiplication *expressions* point to a *relationship* between numbers: a way that these numbers are connected. We will think of an expression as a representation of an underlying relationship. Different expressions may refer to the same relationship. We can use the idea of division to express differently the relationships designated in multiplication expressions.

Figure 5.1 shows three expressions of a relationship between 60, 5, and 12.

We will think about division in other ways also, but this is a simple and convenient approach to step into thinking about division. The key idea here is that the same relationship between three numbers can be expressed with a multiplication or with a division.

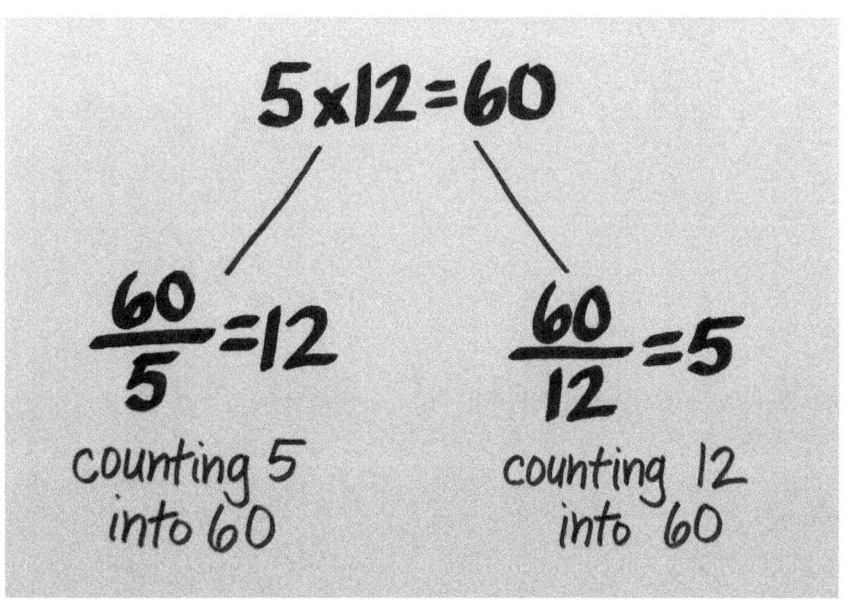

Figure 5.1. Three expressions of the relationship between the three numbers 60, 5, and 12.

The expression

$$4 \cdot 10 = 40$$

connects the numbers 4, 10, and 40. If we look at this relationship from a different perspective—considering first the 40—then we may ask how many times does 10 go into 40? Or *how many times must we count* 10 *to get* 40?

We have a concise notation for this kind of expression using a symbol for division, '/' or '÷'. Expressed in the form of a division we have

$$40 \div 10 = 4$$

and we say *forty divided by ten equals four*.

How many times must we count 10 to get 40? Four times.

This division expression points to exactly the same relationship between 4, 10, and 40, as the prior corresponding multiplication. But we are looking at the relationship from a different perspective. Or you might consider it a different focus. The focus in this division is the counting of 10 into 40. Nevertheless if we know how multiplication works we should be able to use it to figure out divisions.

Example 5.1. Noticing different expressions in the arithmetic relationship between numbers.

5.2 Division Expresses a Counting

We can think of division as a *counting*. For example let's consider

$$12 \div 3 = 4$$

The '÷' sign indicates division.

We read this as twelve divided by three. One way to think about this is as an expression of the counting: *How many times must we count 3 to get 12?* The answer is four. If we count four threes then we get 12. It is common also to write a division in this form:

$$\frac{12}{3} = 4$$

We see the twelve above a bar—the division bar—and the three below that bar. Twelve is in the place of the *numerator* and three is in the place of the *denominator*.

The denominator is the number serving as the counter. The denominator in effect denominates—counts.

The numerator is the number that we are counting with the denominator.

A division expresses the idea: how many times must we count the denominator to get the numerator?

Figure 5.2 illustrates the idea of division as a counting of the denominator into the numerator.

Exercise 5.1. Compute $21 \div 7$

Exercise 5.2. Compute $24 \div 6$

Exercise 5.3. Compute $15 \div 5$

Exercise 5.4. Compute $15 \div 3$

Exercise 5.5. Compute $18 \div 6$

Exercise 5.6. Compute $12 \div 6$

Exercise 5.7. Compute $22 \div 2$

Exercise 5.8. Compute $25 \div 5$

Exercise 5.9. Compute $28 \div 7$

Exercise 5.10. Consider $\frac{20}{4}$. How many times must we count 4 to get 20?

Exercise 5.11. Practice computing divisions using this counting definition. From these exercises you see that multiplication enables the computation of these kinds of simple divisions.

88   CHAPTER 5. DIVISION

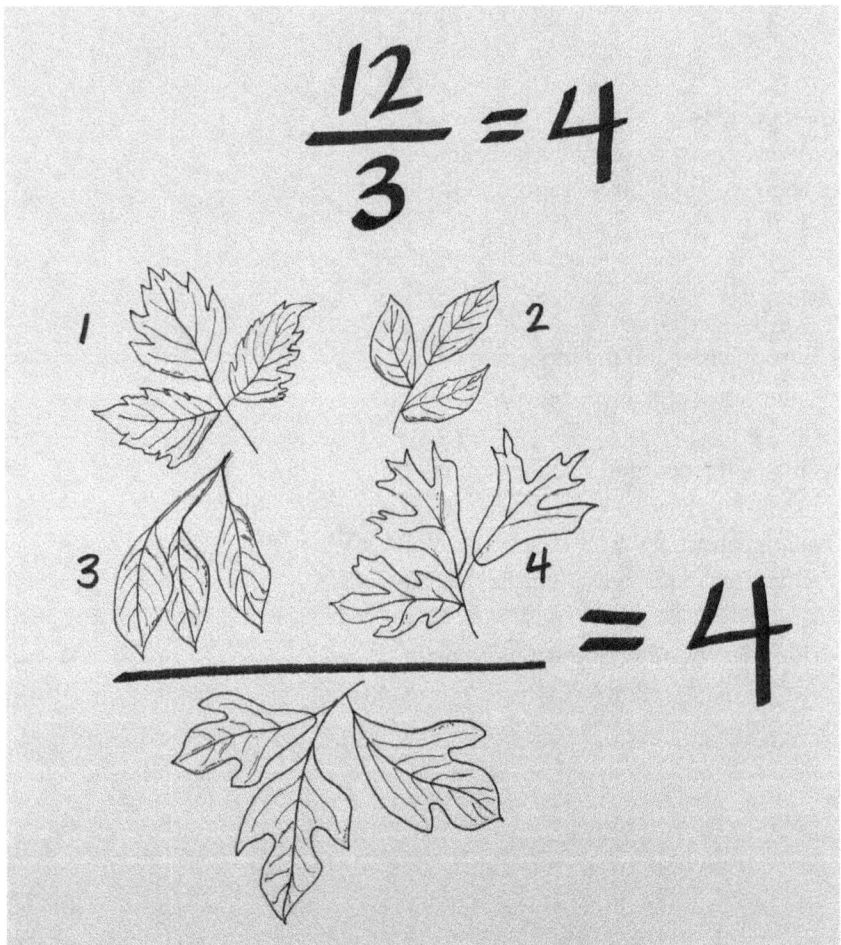

Figure 5.2. We count the denominator, 3, into the numerator, 12. The division shows that four threes make twelve. The groups of leaves illustrate this division also. The division answers the question: how many times must we count the denominator to get the numerator?

Think of division as counting up the denominator. How many times must we count eight to get 24?

If we count three eights, we get 24. This is expressed by the division

$$24 \div 8 = 3$$

Example 5.2. With division we can express the idea of counting eights in 24.

5.2.1 Several Expressions For One Relationship

Going forward when we encounter a multiplication or a division expression we should understand that it points to a relationship that we may reference with several different expressions.

For example

$$60 \div 5 = 12$$

When we know this expression, then we also know the expressions

$$5 \times 12 = 60$$
$$12 \times 5 = 60 \quad \text{and}$$
$$60 \div 12 = 5$$

Example 5.3. The relationship between 60, 12, and 5.

Exercise 5.12. Take a few multiplication or division expressions and think of some of the other expressions between the same numbers. Begin with $100 \div 4 = 25$ and then repeat this exercise with other expressions as well.

Solution:
$4 \cdot 25 = 100$
$100 \div 25 = 4$

(To compute the division $100 \div 4 = 25$, note that 20 fours count up to 80. We then have 20 more to count. Five fours count 20. So $20 + 5 = 25$ fours to count 100.)

Exercise 5.13. Given $40 \div 2 = 20$, use this to find the answer to $46 \div 2$.

Exercise 5.14. Similarly given that $30 \div 3 = 10$ find the answer to $36 \div 3$.

Exercise 5.15. Write the corresponding divisions associated with this multiplication. $1 \cdot 10 = 10$.

Solution:
We can count one ten into ten:

$$10 \div 10 = 1$$

We can also count with one. Ten ones equal ten.

$$10 \div 1 = 10$$

Exercise 5.16. Write the corresponding divisions associated with this multiplication. $2 \cdot 10 = 20$.

Solution:

We can count two tens into twenty.

$$20 \div 10 = 2$$

We can also count ten twos into twenty.

$$20 \div 2 = 10$$

Exercise 5.17. Write the corresponding divisions associated with this multiplication. $3 \cdot 10 = 30$.

Solution:

We count 10 into 30 three times.

$$30 \div 10 = 3$$

We can also count 3 into 30 ten times.

$$30 \div 3 = 10$$

Exercise 5.18. Write the corresponding divisions associated with this multiplication. $11 \cdot 10 = 110$.

Solution:

Eleven goes 10 times into 110

$$110 \div 11 = 10$$

We can count eleven tens into 110.

$$110 \div 10 = 11$$

Exercise 5.19. Write the corresponding multiplication associated with this division. $18 \div 6 = 3$.

Exercise 5.20. Write the corresponding multiplication associated with this division. $20 \div 4 = 5$.

Exercise 5.21. Write five multiplications and then work out the corresponding division expressions.

Exercise 5.22. Choose a number n. Count out a few groups of n. Express the corresponding divisions. For example, if you choose $n = 8$, then you might have three eights are 24, four eights are 32, five eights are 40 and so on.

$$3 \cdot 8 = 24 \quad \Rightarrow \quad 24 \div 8 = 3 \quad \text{and} \quad 24 \div 3 = 8$$
$$4 \cdot 8 = 32 \quad \Rightarrow \quad 32 \div 8 = 4 \quad \text{and} \quad 32 \div 4 = 8$$
$$5 \cdot 8 = 40 \quad \Rightarrow \quad 40 \div 8 = 5 \quad \text{and} \quad 40 \div 5 = 8$$

5.2.2 Visualizing Division

Another useful way to think about division is to imagine a line or a surface or shape that we will cut up or divide into equal parts. This is something we see frequently in building things, measuring things, or in drawing.

For example, we might have a circle divided into two equal parts. Two halves. Figure 5.3 illustrates this.

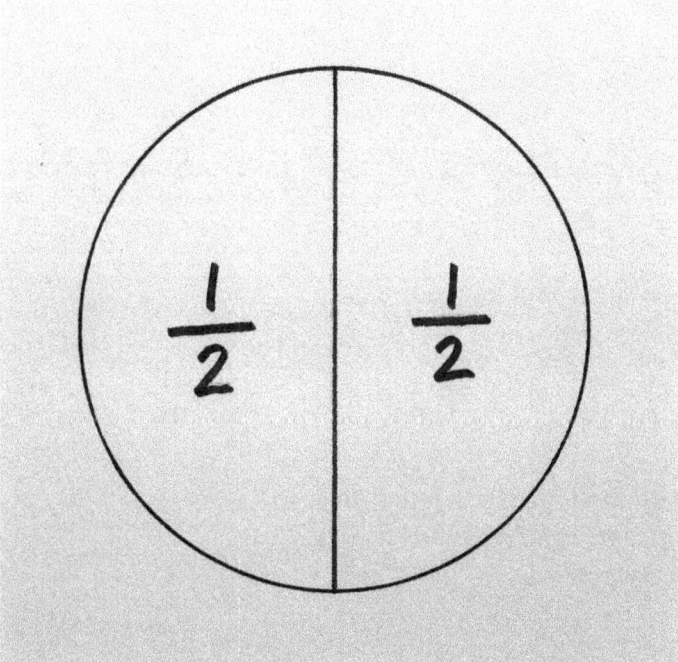

Figure 5.3. Two halves of a circle.

If the area of the whole disc was 10 then the area of each half disc is 5. Figure 5.4 shows some other objects divided into halves.

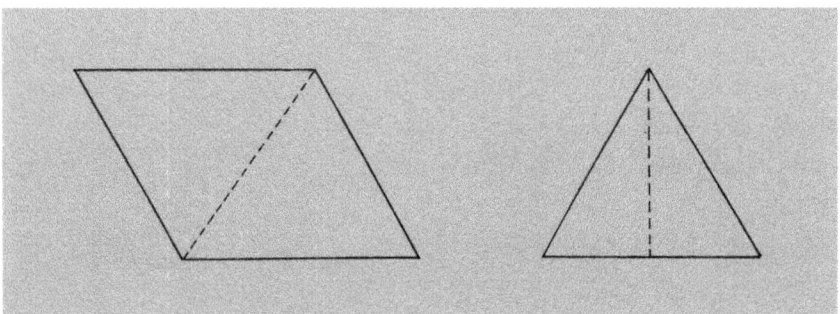

Figure 5.4. A parallelogram and a triangle divided in halves.

5.2.3 Division by One

Let's look now at a few specific cases of division. We will start with a very simple division that must be perfectly clear. Division by one.

For example what is

$$8 \div 1$$

Returning to our counting definition for division, we see that this division would mean *how many ones must we count to get eight?* Counting ones is straightforward. To get eight, we must count eight ones.

More generally for any number n,

$$n \div 1 = n$$

or

$$\frac{n}{1} = n$$

Any time we divide a number n by 1, the answer is n! This is clear and simple from thinking through the counting definition.

Exercise 5.23. Specify $5 \div 1$

Exercise 5.24. Specify $2 \div 1$

Exercise 5.25. Specify $101 \div 1$

5.2.4 *n Divided by n*

Any number divided by itself equals one. This flows easily also from the counting definition. *How many times must we count a quantity n to get n?* One time.

For example, how many times must we count eight to get eight? One eight counts eight. This is expressed in the division

$$\frac{8}{8} = 1$$

More generally for any number *n*

$$n \div n = 1$$

Exercise 5.26. Specify $24 \div 24$

Exercise 5.27. Specify $3 \div 3$

Exercise 5.28. Specify $9 \div 9$

5.2.5 Division by Zero

What about $10 \div 0$?

If we refer again to our counting definition we would ask, *how many times must we count zero to get ten?* How would we make sense of this question? Each time we count zero, we would count nothing. The counting definition of division does not define a division by zero. Division by zero is *not* defined.

Since we don't count very much by counting nothing, we might approach this question another way. Instead of counting nothing, we might try to count something very small but not quite nothing. The smaller the quantity that we count, the more of it we need to count. Observing that trend, we might say that as we approach division by zero, the result of the division gets bigger and bigger. We will look much more at division by very small numbers subsequently.

5.2.6 Division by 2

Exercise 5.29. Compute $\frac{10}{2}$

Exercise 5.30. Compute $\frac{100}{2}$

Exercise 5.31. Compute $\frac{1000}{2}$

Exercise 5.32. Compute $\frac{12}{2}$

Exercise 5.33. Compute $\frac{14}{2}$

Exercise 5.34. Compute $\frac{16}{2}$

Exercise 5.35. Compute $\frac{18}{2}$

Exercise 5.36. Compute $\frac{20}{2}$

Exercise 5.37. Compute $\frac{30}{2}$

Exercise 5.38. Compute $\frac{40}{2}$

5.2.7 Division by 10

Exercise 5.39. Compute $\frac{100}{10}$

Exercise 5.40. Compute $\frac{10}{10}$

Exercise 5.41. Compute $\frac{30}{10}$

Exercise 5.42. Compute $\frac{400}{10}$

Exercise 5.43. Compute $\frac{1000}{10}$

Exercise 5.44. Compute $\frac{1010}{10}$

5.3 Longer Division

Recall that we approached longer multiplications by breaking them down into a sum of smaller multiplications.

$$15 \cdot 12 = 15 \cdot 10 + 15 \cdot 2$$
$$= 150 + 30$$
$$= 180$$

The same strategy works also for division.

$$72 \div 2 = (70 + 2) \div 2$$
$$= 70 \div 2 + 2 \div 2$$
$$= (7 \cdot 10) \div 2 + 1$$
$$= 7 \cdot (10 \div 2) + 1$$
$$= 7 \cdot 5 + 1$$
$$= 35 + 1$$
$$= 36$$

You can think about this again in terms of counting. Imagine a bag containing the 72. You can imagine the 72 in two parts, 70 + 2, inside the bag. So dividing the whole bag in two would mean dividing the 70 by two and also dividing the two by two. This is expressed arithmetically as above.

Let's look at these steps.

The first step simply reflects the meaning of the number 72. Seven tens, or seventy, plus 2.

Next to resolve $72 \div 2$, that is finding half of 72, we can think of 72 as $70 + 2$ and as we want a half of the whole then we take a half of each part. That is $70 \div 2$ and $2 \div 2$.

Two divided by two is simply one. So we focus on $70 \div 2$. Seventy means seven tens. We want half of seven tens.

Seven tens is the same number as ten sevens. Half of ten sevens means of course five sevens. We see we could get this result formally by calculating either $(7 \div 2) \cdot 10$ or $7 \cdot (10 \div 2)$, the parentheses indicating which calculation to do first. Why? Because to compute half of $n \cdot m$ we can either start with half of n and count m of these or we can start with half of m and count them up n times.

At this point we have $72 \div 2 = 7 \cdot 5 + 1$ and we easily get the answer 36.

Exercise 5.45. Compute $\frac{56}{8}$

Solution:

$$\frac{56}{8} = \frac{24}{8} + \frac{32}{8}$$
$$= 3 + 4$$
$$= 7$$

Exercise 5.46. Compute $\frac{112}{8}$

Exercise 5.47. Compute $\frac{72}{9}$

Exercise 5.48. Compute $\frac{153}{9}$

Exercise 5.49. Compute $\frac{48}{6}$

Exercise 5.50. Compute $\frac{48}{8}$

5.4 Division with Remainder

Recall the example where we counted up 24 using eights.

$$\frac{24}{8} = 3$$

What if we want instead to count 25 using eights?

$$\frac{25}{8}$$

We already know that three eights get us to 24. We are almost there. But we still need 1 more to get 25. In other words, having counted three eights for 24, there is still a quantity remaining to get the full count we wish, 25. The remaining quantity, the *remainder*, here is 1.

$$\frac{25}{8} = 3 \quad \text{and a remainder of 1}$$

Let's focus on this remainder of 1. In this division, the denominator 8 tells us that we are counting by eights. How do we count the quantity 1 with the quantity 8? Another way to say this is—what part of 8 is the quantity 1?

We need eight ones to count 8. So a single one is one eighth of eight

$$1 = \frac{1}{8} \cdot 8$$

Now we can rewrite the division $25 \div 8$ more concisely

$$\frac{25}{8} = 3 + 1 \cdot \frac{1}{8}$$

This is an important statement. The remainder shows us the part (or fraction) of the denominator that we must add to the count to complete the quantity that is the division. The remainder always designates a quantity that is smaller than 1. Expressed as a remainder, the number is the numerator in a division with denominator that is the same as the denominator. Another way to see this is to notice:

$$\frac{25}{8} = \frac{24}{8} + \frac{1}{8}$$

This is consistent with our counting definition for division. What's happening here is that we are counting a number that is not a multiple of the denominator. This means that we must count the denominator a whole number of times and then also add a part of the denominator to obtain the complete count of the numerator.

For example in the case of $25 \div 8$ the remainder of 1 tells us

$$\frac{25}{8} = 3 + 1 \cdot \frac{1}{8}$$

This is the same as

$$25 = 3 \cdot 8 + 8 \cdot \frac{1}{8}$$
$$= 24 + 1$$

With the remainder, we can get the complete count of the numerator 25. What about $\frac{26}{8}$.

$$\frac{26}{8} = 3 + 2 \cdot \frac{1}{8}$$
$$= 3 + \frac{1}{4}$$

For the division 26 ÷ 8, we say that the remainder is 2. But it is important to see exactly what the remainder counts. The expression above shows this explicitly. The remainder two indicates that we have two eighths separating the value of the division from the quantity 8×3.

This shows a key idea in division. That is to understand the denominator as indicating a counting unit. So here with denominator 8, a quantity 1 corresponds to $\frac{1}{8}$ of the counting unit. Counting the quantity 1 is equivalent to counting $\frac{1}{8} \cdot 8$. If we need to count a quantity smaller than the counting unit indicated in the denominator, we can simply count the corresponding number of eighths.

Notice also that the question *how should we count* 1 *with* 8*?* leads us to the idea: to count with a part of 8 and in particular with that part of 8 such that 8 of these parts together equal 1.

$$\frac{1}{8} \cdot 8 = 1$$

This key counting idea is also helpful for approaching fractions as we will see later. So your work here will yield many benefits.

Exercise 5.51. Compute $\frac{42}{9}$

Exercise 5.52. $\frac{48}{6}$

Exercise 5.53. $\frac{50}{6}$

Exercise 5.54. $\frac{30}{8}$

Exercise 5.55. $\frac{31}{8}$

Exercise 5.56. $\frac{35}{6}$

Exercise 5.57. $\frac{42}{5}$

Exercise 5.58. $\frac{46}{5}$

Exercise 5.59. $\frac{13}{3}$

Exercise 5.60. $\frac{15}{6}$

Exercise 5.61. $\frac{8}{4}$

Exercise 5.62. $\frac{10}{4}$

Exercise 5.63. $\frac{10}{3}$

Exercise 5.64. $\frac{11}{3}$

Exercise 5.65. $\frac{13}{3}$

Exercise 5.66. $\frac{21}{8}$

Exercise 5.67. $\frac{13}{3}$

Exercise 5.68. $\frac{15}{2}$

Exercise 5.69. $\frac{17}{3}$

Exercise 5.70. $13 \div 7$

Exercise 5.71. $26 \div 5$

Exercise 5.72. $32 \div 6$

Exercise 5.73. $37 \div 9$

Exercise 5.74. $32 \div 10$

Exercise 5.75. $16 \div 4$

Exercise 5.76. $16 \div 5$

Exercise 5.77. $17 \div 3$

Exercise 5.78. $45 \div 6$

Solution:

Seven sixes are 42. $45 - 42 = 3$. That is the remainder. We count six up to 42. And then the remainder is the quantity left, the difference between the numerator 45 and 42, which we counted up to with the denominator 6.

But what is the quotient? It is the whole number counting of the denominator six *and also* the part of six that equals the remainder. The remainder is 3. Three equals $\frac{1}{2}$ of 6. To get 45 we must count 7 full sixes and one $\frac{1}{2}$ of 6.

$45 \div 6 = 7 + \frac{1}{2}$.

Notice that $6 \cdot 7 + 6 \cdot \frac{1}{2} = 42 + \frac{6}{2} = 42 + 3 = 45$. Multiplying the denominator with the quotient is a way to check your division.

Exercise 5.79. $49 \div 8$

Exercise 5.80. $26 \div 6$

Exercise 5.81. $65 \div 7$

Exercise 5.82. $10 \div 3$

5.4.1 A Note on Remainder

In the divisions that we had considered prior to our discussion on remainder, the numerator was always a multiple of the denominator so that we could count to the numerator with a whole number quantity of the denominator. Now that we have developed this concept of remainder, we can handle many more divisions. The key idea is that we count by the denominator so one count is the value of the denominator or we must count some part of the denominator! The unit fraction indicated by the denominator gives us the unit that we can count for parts up to one whole value of the denominator. For example in division by four we are counting fours into the numerator. If we need to count a value smaller than four, there are three options: $\frac{1}{4}$, $2 \times \frac{1}{4}$, $3 \times \frac{1}{4}$.

The remainder concept is very helpful for learning division. It is based in the counting of the denominator into the numerator. And then you can count (or express) the remainder as a part of the denominator.

$$3 \div 3 = 1 \quad \text{(remainder is 0)}$$
$$4 \div 3 = 1 + \frac{1}{3} \quad \text{(remainder is 1)}$$
$$5 \div 3 = 1 + \frac{2}{3} \quad \text{(remainder is 2)}$$
$$6 \div 3 = 2 \quad \text{(remainder is 0)}$$

Example 5.4. The idea of remainder gives us a way to view numbers as organized in a cycle structure. Any whole number divided by three has remainder 0, 1, or 2.

In example 5.4, notice that the quotient increases by one third at each step as we increase the numerator by one. Notice also that the remainder corresponds to a counting of thirds (the denominator as unit fraction). The remainder is 0 or 1 or 2. Here, never more than 2 because the denominator is 3. Why is that?

Of course, if the remainder were 3, we could count another denominator into the the numerator.

5.4.2 Peeling Digits with Division and Remainder

We have seen in our decimal notation system, we express a quantity with a list of numbers formed of digits where each digit counts a power of ten. For example, the number $67,436$ counts 6 ten thousands, 7 thousands, 4 hundreds, 3 tens, and 6 ones. The number is the sum of these counts.

Notice that the concept of remainder gives us a simple way to peel off digits one by one. If we divide by 10, the remainder gives us the ones count. So, the remainder in successive divisions by 10 peel off the digits one after the other.

Figure 5.5 illustrates this idea.

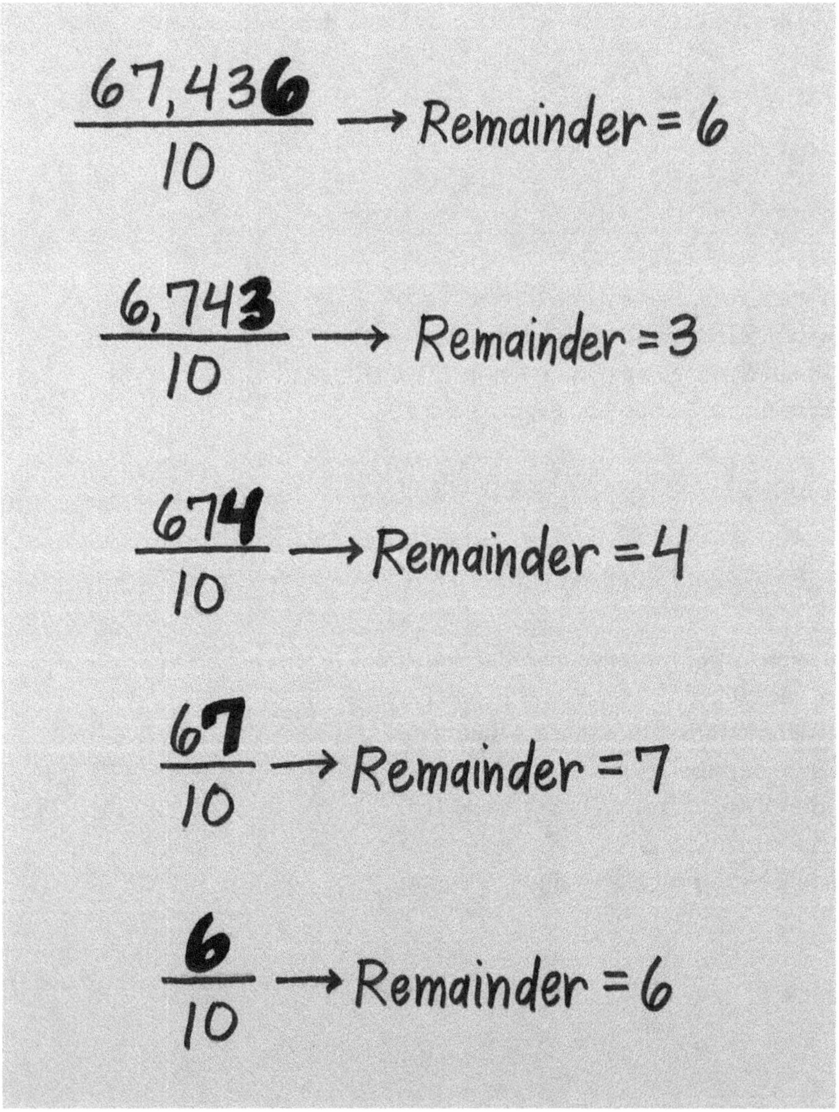

$$\frac{67{,}436}{10} \rightarrow \text{Remainder} = 6$$

$$\frac{6{,}743}{10} \rightarrow \text{Remainder} = 3$$

$$\frac{674}{10} \rightarrow \text{Remainder} = 4$$

$$\frac{67}{10} \rightarrow \text{Remainder} = 7$$

$$\frac{6}{10} \rightarrow \text{Remainder} = 6$$

Figure 5.5. We can peel off the digits of a number using the remainder concept and successive divisions by 10.

5.5 Rational Numbers

Having seen how to divide any two whole numbers and recognizing the meaning of these divisions, we now step back to notice that we have encountered a kind of quantity that we had not been working with previously. Until this discussion, all the quantities that we worked with could be counted up or built by accumulating the quantity 1. We did not need to call on a quantity that was smaller than 1 in forming and representing those quantities. Here however we see quantities that require us to take *a part* of the quantity 1 to build. For example the division $3 \div 2$ requires us to consider the quantity one half—a part of 1—in addition to the 1.

This recognition requires a big and important step in our extension of numbers. Because this extension of whole numbers is so useful we will give it a name. We call *rational numbers* all the numbers that we can build from the division of two whole numbers. Rational numbers have the form

$$\frac{a}{b} \quad \text{for } a \text{ and } b \text{ whole numbers}$$

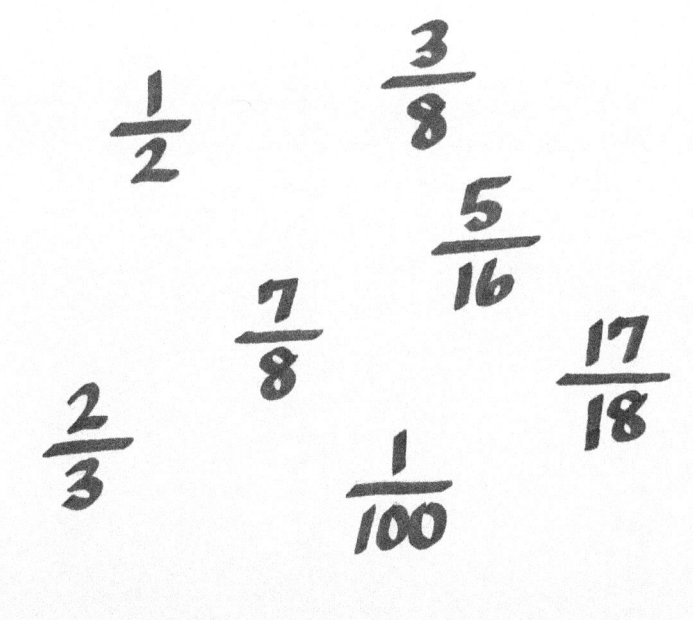

Figure 5.6. Illustrating various rational numbers of the form $\frac{a}{b}$ for a and b whole numbers.

Notice that all the whole numbers we were familiar with previously are also included in our new collection of rational numbers because any whole number can also be expressed as a division. We saw this in section 5.2.3. Here are more examples.

$$2 = \frac{2}{1}$$
$$3 = \frac{3}{1}$$
$$10 = \frac{10}{1}$$
$$1000 = \frac{1000}{1}$$
$$4,321 = \frac{4,321}{1}$$

In fact there are actually infinitely many ways to represent a whole number in *rational form* (that is, as a division of two whole numbers).

$$3 = \frac{6}{2}$$
$$= \frac{9}{3}$$
$$= \frac{30}{10}$$
$$= \frac{33}{11}$$
$$= \frac{300}{100}$$

Exercise 5.83. Choose 10 different whole numbers. For each of these numbers, write five corresponding rational forms. For example, if I choose the whole number 15, I might write

$$15 = \frac{15}{1}$$
$$= \frac{30}{2}$$
$$= \frac{45}{3}$$
$$= \frac{150}{10}$$
$$= \frac{300}{20}$$

5.5.1 Unit Fractions

Now that we have developed this idea of rational numbers, we can use this to think about division in a different way. Rather than as a counting of the denominator into the numerator we can think of division as a multiplication: a counting of a unit fraction.

5.5. RATIONAL NUMBERS

A *unit fraction* is simply a rational number where the numerator is 1. For example $\frac{1}{2}, \frac{1}{3}, \frac{1}{4}, \frac{1}{10}, \frac{1}{100}$, these are unit fractions. More generally, a unit fraction is any number with the form

$$\frac{1}{n} \quad \text{where } n \text{ is a whole number}$$

Notice now that a rational number, that is a number with the form $\frac{a}{b}$ can be expressed as a multiplication of a unit fraction:

$$\frac{a}{b} = a \cdot \frac{1}{b}$$

We can think of the rational number $\frac{a}{b}$ as a counting of the unit fraction $\frac{1}{b}$. $\frac{a}{b}$ counts a times the unit fraction $\frac{1}{b}$.

Exercise 5.84. Write the following division as a multiplication of a unit fraction. $\frac{4}{5}$.

Exercise 5.85. Write the following division as a multiplication of a unit fraction. $\frac{4}{10}$.

Exercise 5.86. Write the following division as a multiplication of a unit fraction. $\frac{4}{100}$.

Exercise 5.87. Write the following division as a multiplication of a unit fraction. $\frac{4}{1000}$.

Exercise 5.88. Write the following division as a multiplication of a unit fraction. $\frac{4}{10000}$.

Exercise 5.89. Write the following division as a multiplication of a unit fraction. $\frac{3}{4}$.

Exercise 5.90. Write the following division as a multiplication of a unit fraction. $\frac{3}{8}$.

Exercise 5.91. Write the following division as a multiplication of a unit fraction. $\frac{3}{16}$.

Exercise 5.92. Write the following division as a multiplication of a unit fraction. $\frac{3}{32}$.

Exercise 5.93. Write the following division as a multiplication of a unit fraction. $\frac{13}{52}$.

Exercise 5.94. Write the following division as a multiplication of a unit fraction. $\frac{75}{100}$.

Exercise 5.95. Write the following division as a multiplication of a unit fraction. $\frac{750}{1000}$.

But what is the meaning of a unit fraction? It is the part of 1 resulting when we divide 1 into as many equal parts as indicated by the number in the denominator. For example $\frac{1}{4}$ designates the quantity, four of which add up to 1.

$$\frac{1}{4} + \frac{1}{4} + \frac{1}{4} + \frac{1}{4} = 1$$

Figure 5.7 also illustrates this.

Figure 5.7. The value of one fourth.

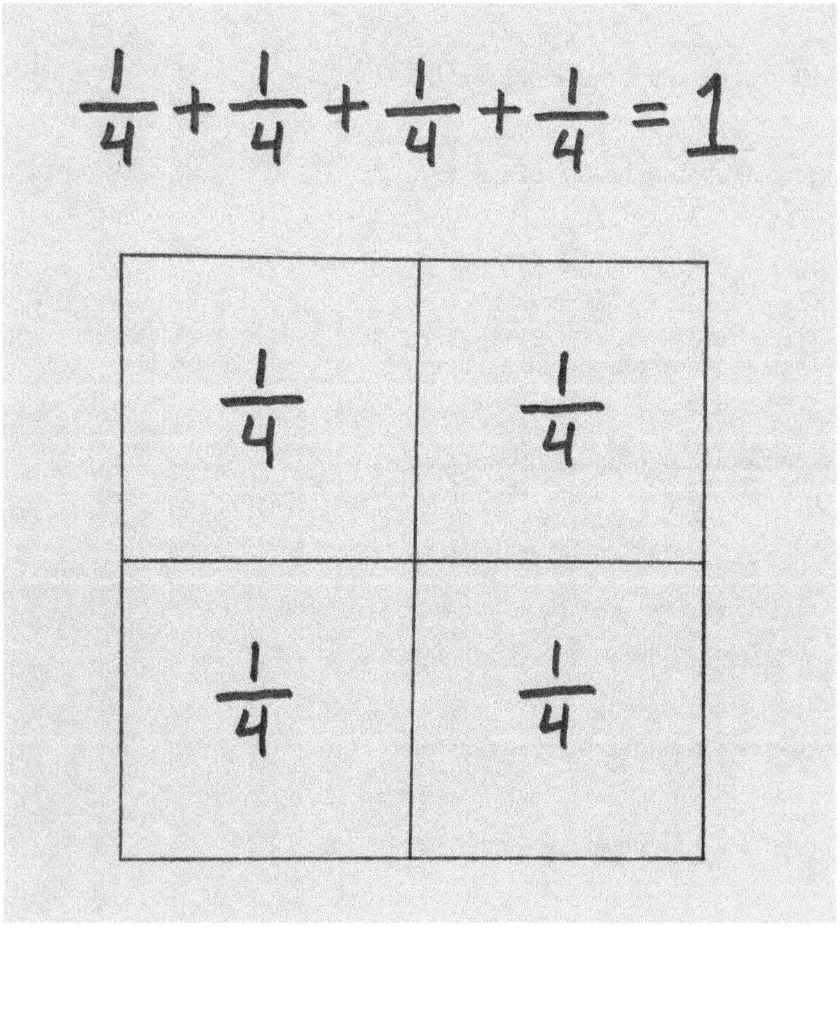

Similarly

$$\frac{1}{8} + \frac{1}{8} + \frac{1}{8} + \frac{1}{8} + \frac{1}{8} + \frac{1}{8} + \frac{1}{8} + \frac{1}{8} = 1$$

We will look at fractions, including unit fractions, in great detail in chapter 7. But it is helpful to have a sketch of this idea handy at this point.

5.5.2 A Quick Look Ahead: Dividing Unit Fractions

What if we divide a unit fraction? That is, we a take a quantity smaller than 1 and we seek, by division, to take a part of that. For example, what if we are interested in the sixth part of one seventh?

$$\frac{\left(\frac{1}{7}\right)}{6}$$

From what we recently saw on unit fractions, we can rewrite this division.

$$\frac{\left(\frac{1}{7}\right)}{6} = \frac{1}{7} \cdot \frac{1}{6}$$

Can you think of how to make sense of this and simplify this expression? In chapter 7, we will discuss in detail how to handle multiplications like this. But see if you can figure it out on your own before then.

5.6 Inverse

With the idea of unit fractions available, we can discuss a very important and useful concept. The inverse under multiplication. This important idea is based in a very simple definition. Two numbers are *inverses* of each other when their multiplication yields 1. For example 3 and $\frac{1}{3}$ are inverses of each other because

$$3 \cdot \frac{1}{3} = 1$$

More generally for any number n: $\frac{1}{n}$ and n are inverses of each other because

$$n \cdot \frac{1}{n} = 1$$

This is a crucial idea for multiplication and also it gives us another perspective into division as we will see subsequently.

Ten times the inverse of ten equals 1. We will write the inverse of ten as $\frac{1}{10}$.

$$10 \times \frac{1}{10} = 1$$

Example 5.5. 10 multiplied with its inverse equals 1.

To determine the inverse of a number, we can ask the question what must we multiply with this number to get 1.

Exercise 5.96. What is the inverse of $\frac{1}{8}$?

Solution:
$8 \cdot \frac{1}{8} = 1$. So 8 is the inverse of $\frac{1}{8}$. Notice that of course $\frac{1}{8}$ is the inverse of 8.

Exercise 5.97. What is the inverse of 10?

Solution:
$10 \cdot \frac{1}{10} = 1$. So the inverse of 10 is $\frac{1}{10}$

Exercise 5.98. What is the inverse of 1?

Exercise 5.99. Can you figure out the inverse of $\frac{3}{4}$? We will discuss this in detail in chapter 7 on fractions, but see if you can find the answer now.

Exercise 5.100. What is the inverse of 0?

Solution:
There is no inverse for 0. You could say that it is not defined. Or you might say that it is infinity. If you took a step off of 0 just a little bit so you have a really small number. Say $\frac{1}{1,000,000}$, that would be the inverse of $1,000,000$. So in this sense you might think of the inverse of 0 as infinity, an ever bigger number. But 0 has no inverse. If you wiggle a little bit away from 0 you get a feel of how the situation is trending near 0.

5.7 Simplification of Rational Numbers

As we build up or work with more complicated expressions it is often convenient to simplify the forms of expressions whenever possible especially rational numbers.

Simplifying rational numbers is based in the fact that a number divided by itself equals 1. For example

$$\frac{5}{5} = 1$$

And generally

$$\frac{n}{n} = 1$$

So whenever we have a rational form including a multiple of a number divided by itself we can replace this multiple with the factor 1.

For example

$$\begin{aligned}\frac{8}{12} &= \frac{4 \cdot 2}{4 \cdot 3} \\ &= \frac{4}{4} \cdot \frac{2}{3} \\ &= 1 \cdot \frac{2}{3}\end{aligned}$$

Very important: we are simplifying a *factor* in the numerator and in the denominator. This must be a factor and not a term (a term is a quantity being added to another). Simplifying a common factor in the numerator and the denominator makes sense. What does it mean, for example to multiply the numerator by 4? We are making the overall value four times bigger. And when we multiply the denominator by four, we are making the overall expression four times smaller. So if we cancel out the common factor of four in the numerator and in the denominator, we are simply avoiding the unnecessary work of making the expression four times bigger and then four times smaller. This does not change the value of the expression.

Figure 5.8 illustrates this process of simplification.

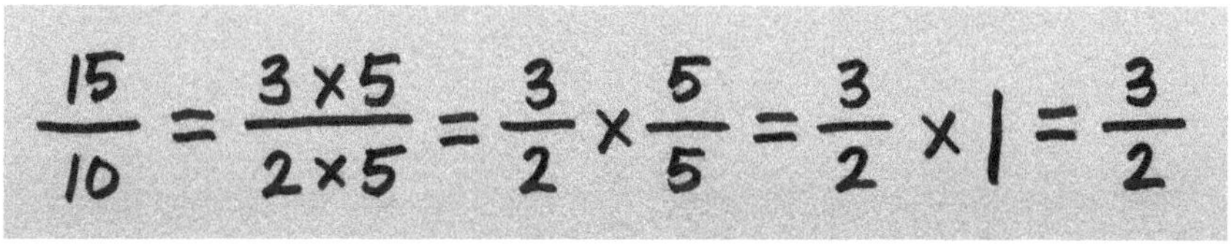

Figure 5.8. Simplifying the rational number $\frac{15}{10}$.

$$\frac{15}{10} = \frac{3 \times 5}{2 \times 5} = \frac{3}{2} \times \frac{5}{5} = \frac{3}{2} \times 1 = \frac{3}{2}$$

Exercise 5.101. Explain why the following is *incorrect*:

$$\frac{7}{6} = \frac{4+3}{4+2} = \frac{3}{2}$$

Solution:

The last expression $\frac{3}{2}$ is *not* equal to $\frac{7}{6}$. It looks like a four was subtracted from the numerator and the denominator. But this changes the value of the expression rather than preserving the value. The last equal sign is false. We can remove from the numerator and denominator a *factor* in common because arithmetically the effect of the common factors combined is to cancel each other. The factor in the numerator is like an amplification. The factor in the denominator is the corresponding shrinking. On the other hand, as a general matter if we add a quantity to the numerator and the denominator we are not preserving the value of the expression over all. This is easy to see in a few examples.

$$\frac{10+1}{5+1} = \frac{11}{6} \neq 2 \quad \text{whereas}$$
$$\frac{10}{5} = 2$$

Or

$$\frac{30+1}{10+1} = \frac{31}{11} \neq 3 \quad \text{whereas}$$
$$\frac{30}{10} = 3$$

Generally, adding a quantity to the numerator and the denominator of a rational number will change the value of that number. In contrast, multiplying the numerator and the denominator with a given quantity will not change the value of that number.

Exercise 5.102. Simplify $\frac{3 \cdot 10}{4 \cdot 10}$

Exercise 5.103. Simplify $\frac{30}{40}$

Exercise 5.104. Simplify $\frac{5 \cdot 6}{5 \cdot 7}$

Exercise 5.105. Simplify $\frac{30}{35}$

Exercise 5.106. Simplify $\frac{9 \cdot 4}{2 \cdot 9}$

Exercise 5.107. Simplify $\frac{36}{18}$

Exercise 5.108. Simplify $\frac{8 \cdot 10}{6 \cdot 8}$

Exercise 5.109. Simplify $\frac{80}{48}$

6 Subtraction & Directed Numbers

6.1 Counting Down

Subtraction flows directly from considering any addition, for example $10 + 8 = 18$. If we start with the quantity 18 and if we remove from it the quantity eight, then we are left with ten. Similarly, if we take 10 from 18, we are left with eight.

$$18 - 8 = 10$$
$$18 - 10 = 8$$

These expressions illustrate *subtraction*. And indeed, one way to think about subtraction is that we are removing (or subtracting) one quantity from another. Subtraction is thus in some sense the opposite of addition. We designate subtraction with the symbol '−'.

$$10 - 8 = 2$$

We find the answer (the *difference*), two, by counting back eight from ten. That is, we remove eight from ten. We have two remaining.

Alternatively, we might have had in mind that $10 = 8 + 2$ and determined the value of $10 - 8$ from that equality.

Subtraction highlights the idea of direction in counting. We can now count up or down (to the right or the left on a horizontal number line, if you prefer). Two is what remains when we take eight away from ten. If we think of the quantity ten as made up of two parts, eight and two, then we see that only two remains if eight is removed.

Example 6.1. We can find the difference of ten minus eight from the related addition eight plus two. Or we can count back from ten.

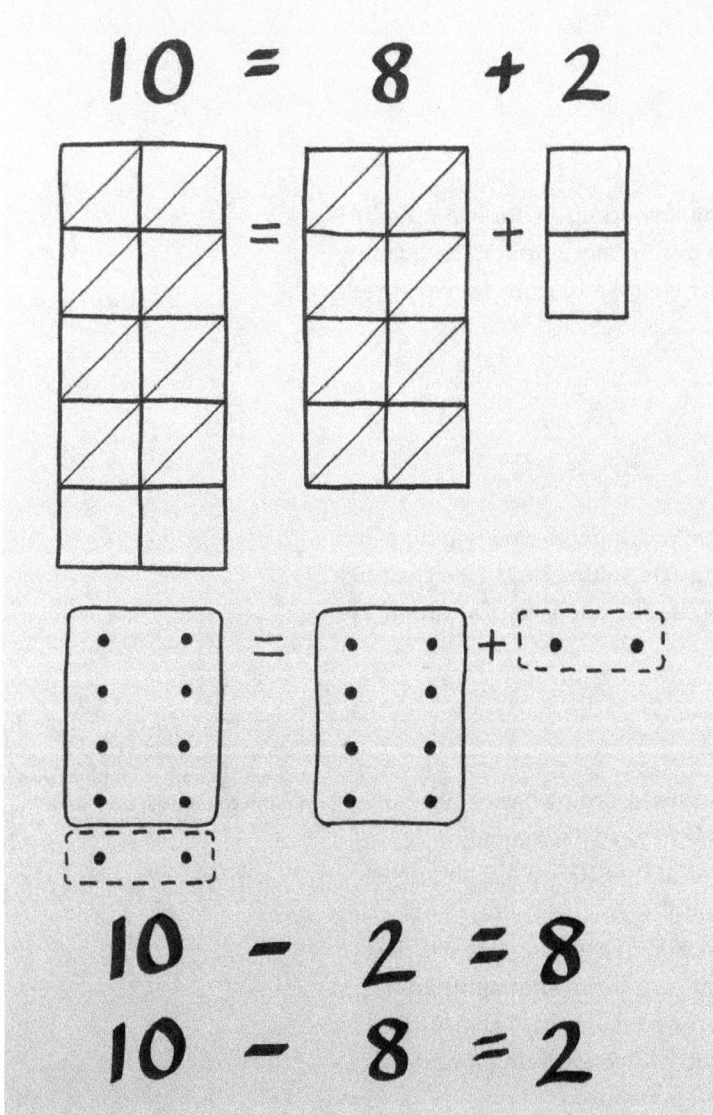

Figure 6.1. Subtraction from the corresponding addition.

Exercise 6.1. Write down *twelve* addition expressions and then for each one write two corresponding subtraction expressions.

Solution:

$$7 + 8 = 15 \implies \begin{cases} 15 - 8 = 7 \\ 15 - 7 = 8 \end{cases}$$

$$3 + 9 = 12 \implies \begin{cases} 12 - 3 = 9 \\ 12 - 9 = 3 \end{cases}$$

Exercise 6.2. Determine $15 - 6$

Exercise 6.3. Determine $9 - 3$

Exercise 6.4. Determine $18 - 7$

Exercise 6.5. Determine $19 - 3$

Exercise 6.6. Determine $19 - 13$

Exercise 6.7. Determine $100 - 50$

Exercise 6.8. Determine $23 - 4$

Exercise 6.9. Determine $27 - 9$

Exercise 6.10. Determine $12 - 4$

Exercise 6.11. Choose *ten* two-digit numbers. For each of these write down three sums and the corresponding subtractions for those sums.

Solution:

For the number 12, we can write

$$12 = 7 + 5$$
$$12 = 10 + 2$$
$$12 = 9 + 3$$

$$12 = 7 + 5 \implies \begin{cases} 12 - 7 = 5 \\ 12 - 5 = 7 \end{cases}$$

$$12 = 10 + 2 \implies \begin{cases} 12 - 2 = 10 \\ 12 - 10 = 2 \end{cases}$$

$$12 = 9 + 3 \implies \begin{cases} 12 - 9 = 3 \\ 12 - 3 = 9 \end{cases}$$

The addition expression and a corresponding subtraction expression describe the same relationship between three quantities just from a different perspective. Consider

$$14 - 8$$

If you find it helpful, you might think of this subtraction as asking the question *what must I add to 8 to get* 14?

6.1.1 Using Round Numbers in Subtraction

It is sometimes helpful to compute subtractions by counting up and using round numbers as stepping stones. Recall, round numbers are numbers that end in zero and are easy to work with. For example, 10, 20, 30, 40, 50 and so on.

$$78 - 49 = ?$$

Let's count up from 49.

We need 1 to get to 50. Then from 50 to 70, that is 20. Finally we need 8 to get to 78. So counting up the gap (the difference) from 49 to 78 we get

$$1 + 20 + 8 = 29$$

We can also check this answer with the addition $49 + 29$.

$$49 + 29 = 78$$

This confirms that we counted the original difference correctly. It is a good practice to check your computations whenever you can.

Let's now try to subtract 49 from 78 directly.

$$78 - 49$$

We can think of this subtraction in stages and there are lots of possibilities for the stages according to what is easiest for you. For example, we might think of 49 as $40 + 9$. So subtracting 49 is the same as subtracting 40 and then subtracting 9.

$$78 - 40 = 38$$

This subtraction is easy because all we have to do is subtract the 4 (counting tens) from the 7 (counting tens) since in 40 there are no ones. To complete the original subtraction we must also subtract the 9.

$$\begin{aligned} 38 - 9 &= 30 + 8 - 9 \\ &= 30 + 8 - 8 - 1 \\ &= 30 - 1 \\ &= 29 \end{aligned}$$

If it is helpful in the following subtractions, think of how to work out the answer in stages.

Exercise 6.12. $89 - 53$

Exercise 6.13. $87 - 49$

Exercise 6.14. $101 - 62$

Exercise 6.15. $37 - 18$

Exercise 6.16. $76 - 67$

Exercise 6.17. $1000 - 102$

Exercise 6.18. $1000 - 99$

Exercise 6.19. $999 - 98$

6.2 The Other Side of Zero

So far we have considered subtracting a smaller quantity from a bigger one. If we flip the order we can consider the case of taking away a bigger quantity from a smaller one. We'll start with a typical subtraction

$$18 - 10 = 8$$

We'll just switch the 18 and the 10.

$$10 - 18$$

Here is the subtraction step by step

$$\begin{aligned} 10 - 18 &= 10 - 10 - 8 \\ &= -8 \end{aligned}$$

We see that after subtracting ten from the original ten, we still have eight remaining to subtract, as indicated by the minus sign in front of the 8. Numbers that are marked with a minus sign like this are called *negative numbers*. They indicate a counting below zero rather than above zero. As we count the negative numbers −1, −2, −3, −4, and so on we move away from zero but in the opposite direction from when we count the *positive numbers*.

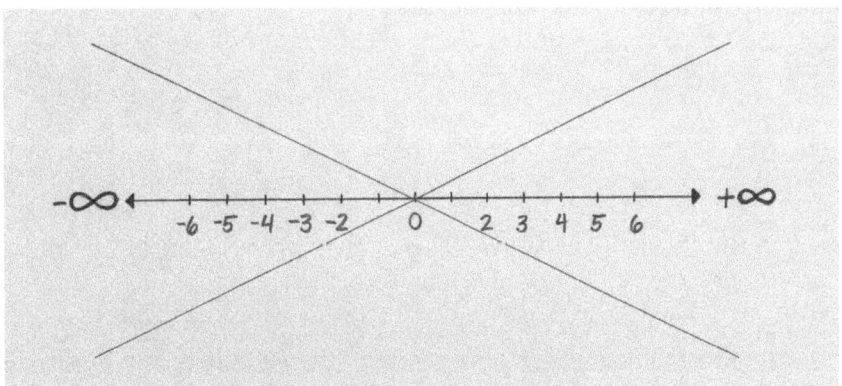

Figure 6.2. −∞ is to the left and +∞ is to the right on this number line.

6.2.1 The Minus Sign Indicates Direction or Orientation

One useful way to think about negative (and positive) numbers is to visualize them on a line marked with numbers as shown in figure 6.2. Let's choose a point in the middle of our line to represent 0, a starting point that we call the *origin*. If we move to the right on this line then the points on the line represent increasing positive numbers. These are numbers greater than zero. Moving to the left corresponds to decreasing numbers. These are negative numbers—numbers less than zero. Notice that the positive and negative numbers are symmetrical about the origin. This means that for every positive number n there is a corresponding negative number $-n$. And n and $-n$ are the same distance from the origin 0. This distance from 0 to n or from 0 to $-n$ is important enough that we give it a name: this distance is called the *absolute value of n*. We also write the absolute value of n concisely as $|n|$.

What are these negative numbers? What can a negative quantity be? We actually use them all the time. They are useful to describe many relationships. For example we use negative numbers to express temperatures. Our common units for temperature—Fahrenheit and Celsius—both range into positive and negative numbers for temperatures that we experience. The Celsius units are based on water properties. 0 degrees Celsius corresponds to the temperature at which water freezes at sea level atmospheric pressure. But we frequently experience temperatures colder than this. At the North Pole the average winter temperature is -40 degrees Celsius. We might also use negative numbers in measuring ocean depth from sea level for example if we want to have sea level correspond to a height of zero with positive values referring to height above sea level and negative values referring to depth. The Marianas Trench in the Pacific Ocean is about thirty-five thousand feet deep. We might write that as $-35,000$ feet to express the idea that the trench is $35,000$ feet *below* the surface of the sea. That is as far below the sea as you might see a passenger jet flying overhead.

Notice that once again by exploring the scope of an arithmetic operation, here subtraction, we have identified a new kind of number: the negative numbers. From subtraction, we see the need to distinguish negative and positive numbers depending on whether we are describing a value that is less than or greater than zero.

Subtraction also leads us to another useful way that we can characterize numbers. We can ascribe to a number a sign that indicates the direction of counting. Positive numbers are counted up from zero. Negative numbers are counted down from zero. On a number line usually the negative numbers are to the left of zero and the positive numbers are to the right of zero. Notice now we can use this new characteristic of number to designate *direction* as well as *magnitude* (bigness). Moving along a line, there are two directions from a point such as the origin zero. So we need only two signs to distinguish the two options that we have for direction. We use the minus sign and the plus sign for this. See figure 6.3 illustrating how the plus and minus signs indicate direction.

When we move towards bigger and bigger numbers we can speak of moving in the direction of positive infinity ($+\infty$). We can also move in the direction of negative infinity ($-\infty$) using subtraction. For example the numbers -10, -100, $-1,000$, $-10,000$, $-100,000$ show a sequence of numbers advancing towards $-\infty$. Notice that although the ideas of $+\infty$ and $-\infty$ do not denote specific numbers themselves, they are useful here as references to two opposite directions in which we can move with a sequence of numbers.

6.2.2 Symmetry About Zero

The number line spanning from $-\infty$ to $+\infty$ is symmetric about the origin, zero (see figure 6.3). This means that the gap—or *difference*—between two positive numbers n and m is the same as the gap separating the two negative numbers $-n$ and $-m$.

116   CHAPTER 6. SUBTRACTION & DIRECTED NUMBERS

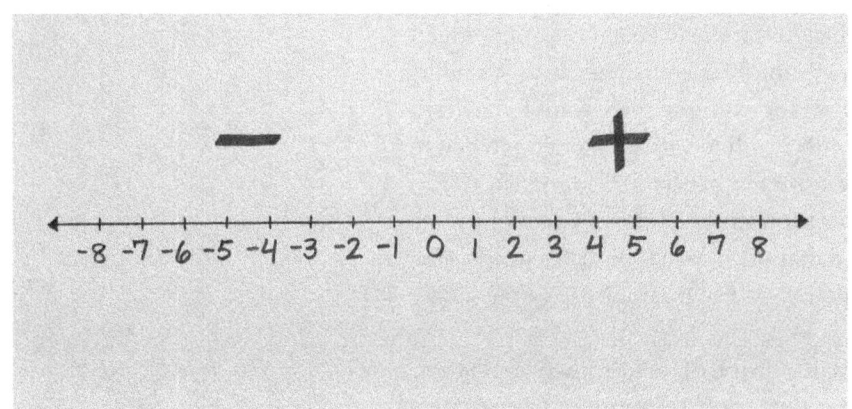

Figure 6.3. We use the plus sign and the minus sign to designate direction on the number line.

For example, the difference separating 8 and 18 is 10. $8 + 10 = 18$ or $18 - 10 = 8$. Eight and 18 are separated by 10 ones.

Similarly $-8 - 10 = -18$ or $-18 + 10 = -8$. The difference separating -18 and -8 is also 10 ones.

Note also that

$$8 - 8 = 0$$

$$10 - 10 = 0$$

$$18 - 18 = 0$$

That is, 0 is half way between n and $-n$.

Exercise 6.20. Compute $-10 - 10$

Solution:

$-10 - 10 = -20$. Minus ten minus ten equals minus twenty. If we take ten away from minus ten we go ten deeper into the negative numbers. We go to -20.

Exercise 6.21. Compute $10 - 20$

Solution:

$10 - 20 = -10$. Ten minus twenty equals minus ten.

Exercise 6.22. Compute $-18 + 10$

Solution:
$-18 + 10 = -8$

Exercise 6.23. Compute $-18 + 20$

Solution:
$-18 + 2 = 2$. (Or $+2$.)

6.2.3 Absolute Value

We now have negative and positive numbers to work with. For every positive number n, we can express a negative number, $-n$, simply by writing the minus sign in front of n.

There are two parts to our number representation here. The magnitude of the number without the sign. This is called the *absolute value* of the number. And the sign $+$ or $-$. If the sign of a number is positive, we usually do not write the $+$ and conventionally if a number has no sign it is assumed to be positive. The absolute value of n is referenced as $|n|$. The absolute value notation enables us to designate the magnitude of a number without its sign.

$$|-8| = 8$$

$$|+8| = 8$$

We reference magnitude or absolute value when we are interested in the size of the number but we don't care about direction or the attribute described by the sign.

6.3 Arithmetic with Subtraction & Negative Numbers

From what we have just discussed, one way to think about the negative numbers is that they are like a *mirror image* of the positive numbers: a reflection across the origin of the number line. And if we are comfortable working with the positive numbers, this should help us think through working with the negative numbers also. In this section we will look more carefully at arithmetic with negative numbers.

In extending our understanding of number to include negative numbers we are just adding to number the characteristic of *direction*. Prior to this, we were working with *magnitudes*. Now we have not only magnitude but also direction. This means that we can have two numbers of the same magnitude but of opposite direction. For example, the numbers 18 and -18. These numbers both have the same magnitude. Their magnitude is 18. But their directions are opposite. 18 counts in the positive direction. -18 counts in the negative direction.

118   CHAPTER 6. SUBTRACTION & DIRECTED NUMBERS

Appreciating both these characteristics of number—magnitude and direction—will help us work easily with negative numbers. All we have to do is to track both of these characteristics of number as we work through operations.

Exercise 6.24. $-2 + -3$

Solution: This is the mirror image of adding $2 + 3$. We can also think of this as simply subtracting 3 from -2. This takes us to -5. Notice that adding a negative number can be thought of equivalently as subtracting a positive number.

$$-2 + -3 = -2 - 3 = -5$$

Exercise 6.25. $-12 + -8$

Exercise 6.26. $-5 + -3$

Exercise 6.27. $-23 + -14$

Exercise 6.28. $-20 + -80$

Exercise 6.29. $-36 + -4$

Exercise 6.30. $-21 + -12$

Exercise 6.31. $-13 + -8$

Exercise 6.32. $13 + -8$

Solution: This can be rewritten as a simple subtraction

$$13 + -8 = 13 - 8 = 5$$

Exercise 6.33. $23 + -8$

Exercise 6.34. $13 + -14$

Exercise 6.35. $13 + -15$

Solution: When we are subtracting a bigger from a smaller quantity, the answer will be a negative number. We can think of this in two steps. A first step subtracting the same quantity to count down to zero. Then what remains to subtract shows the difference.

$$13 + -15 = 13 - 15 = 13 - 13 - 2 = 0 - 2 = -2$$

Exercise 6.36. $12 + -25$

Exercise 6.37. $55 + -62$

Exercise 6.38. $99 + -100$

Exercise 6.39. $46 + -97$

Exercise 6.40. $3 + -27$

Considering multiplication gives us a very useful way to think about negative numbers. Multiplication provides an expression that clearly delineates the two key characteristics of these numbers: direction (sign) and magnitude (absolute value). We have two designations for sign $+$ and $-$ although the absence of sign signifies $+$. As we have seen, the absolute value of a number specifies magnitude.

With multiplication, a negative number n may be thought of as $-1 \cdot |n|$. For example

$$-8 = -1 \cdot 8$$
$$18 = +1 \cdot 18$$
$$+3 = +1 \cdot 3$$
$$-2 = -1 \cdot 2$$

When we multiply with negative numbers, we can work through the operation in two steps. First, to determine the sign. And then to determine the value. Moreover we can determine the sign of the multiplication if we understand the simple logic that is set out in the multiplications between 1 and -1. There are four possible multiplications and one of these is the trivial $1 \cdot 1 = 1$.

Here are the other three to know fluently:

$$-1 \cdot (-1) = 1$$
$$1 \cdot (-1) = -1$$
$$-1 \cdot 1 = -1$$

An analogy that might provide some insight is to think of multiplication by -1 as flipping a light switch as far as sign is concerned. The light is either on or off. If you flip the switch once, you change the state of the light. But if you flip the switch twice then you are back in the original state. If your light is on, and you flip the switch twice, you end up still with the light on.

These simple facts make multiplication and division with negative numbers easy. Determine the absolute value of the answer by operating on the absolute values. The sign of the answer is negative if there is an odd number of minus signs and positive otherwise.

$$---1 = -1$$
$$----1 = 1$$

Notice, by the way, as a matter of notation, that $---1 = -1 \cdot -1 \cdot -1$. Each minus sign indicates multiplication by -1.

Exercise 6.41. Compute $2 \cdot (-3)$

Solution:
$2 \cdot (-3) = (1 \cdot (-1)) \cdot (2 \cdot 3) = -1 \cdot 6 = -6$. We can separate out the determination of the sign for the product and the determination of its absolute value. The sign is the product of the signs. The absolute value is the product of the absolute values.

Exercise 6.42. Compute $-3 \cdot (-4)$

Solution:
$-3 \cdot -4 = (-1 \cdot -1) \cdot (3 \cdot 4) = 12$

Exercise 6.43. Compute $6 \cdot (-5)$

Exercise 6.44. Compute $-12 \cdot (4)$

Exercise 6.45. Compute $8 \cdot (-6)$

Exercise 6.46. Compute $7 \cdot (-5)$

Exercise 6.47. Compute $6 \cdot (-13)$

Exercise 6.48. Compute $-10 \cdot (-10)$

Exercise 6.49. Compute $-5 \cdot (-5)$

Exercise 6.50. Compute $-5 \cdot (-5) \cdot (-5)$

Solution:
$-5 \cdot (-5) \cdot (-5) = (-5 \cdot (-5)) \cdot (-5) = 25 \cdot (-5) = -125$

Exercise 6.51. Compute $-4 \cdot (-4) \cdot (-4)$

Exercise 6.52. Compute $-3 \cdot (-3) \cdot (-3)$

Exercise 6.53. Compute $-2 \cdot (-2) \cdot (-2)$

Exercise 6.54. Compute $-1 \cdot (-1) \cdot (-1)$

Exercise 6.55. Compute $5 \cdot (5) \cdot (5)$

Exercise 6.56. Compute $-10 \cdot (-10) \cdot (-10)$

Exercise 6.57. Compute $-1 \cdot 2 \cdot (-3)$

Exercise 6.58. Compute $2 \cdot (-3) \cdot 4$

Exercise 6.59. Compute $-3 \cdot 4 \cdot (-5)$

As a negative sign can be expressed with the factor of (-1), division with negative numbers is easy if multiplication with negative numbers is understood. The same approach works. We can determine the sign and the absolute value in separate steps. And the same method applies for determining the sign.

$$\frac{-1}{-1} = +1$$
$$\frac{-1}{1} = -1$$
$$\frac{1}{-1} = -1$$

Also note that

$$-\frac{-1}{-1} = -1$$

Exercise 6.60. Compute $\frac{12}{-4}$

Solution:
$\frac{12}{-4} = -\frac{12}{4} = -3$

Exercise 6.61. Compute $\frac{-12}{-4}$

Solution:
$\frac{-12}{-4} = +\frac{12}{4} = 3$

Exercise 6.62. Compute $-\frac{-12}{-4}$

Solution:
$-\frac{-12}{-4} = -\frac{12}{4} = -3$

Exercise 6.63. Compute $\frac{36}{-9}$

Exercise 6.64. Compute $\frac{-56}{-7}$

Exercise 6.65. Compute $\frac{-72}{8}$

Exercise 6.66. Compute $\frac{-144}{-12}$

Exercise 6.67. Compute $\frac{-16}{4}$

6.3.1 Exercises. Subtracting with Negative Numbers.

Exercise 6.68. Compute $--8 = 8$

Solution:
$--8 = -1 \cdot (-1 \cdot 8) = 8$

Exercise 6.69. Compute $12 - -4$

Exercise 6.70. Compute $-12 - -4$

Exercise 6.71. Compute $8 - -18$

Exercise 6.72. Compute $2 - -4$

Exercise 6.73. Compute $0 - -1$

7 Fractions

7.1 A Part of One

The term *fraction* literally means a part—a part of a whole; a part of one. It derives from the Latin *fractio* meaning *breaking*.

The shift in perspective for this discussion is from counting ones and building whole number quantities to considering what we can do if we break one into pieces and we count those pieces.

When we were working only with whole numbers in our additions, subtractions, and multiplications, those operations always yielded also whole numbers. So if we limited our work to those operations with whole numbers we never encountered a different kind of number. That changes when we start working with division.

Exercise 7.1. Can you think of a number that is not a whole number?

Let's consider the division

$$\frac{2}{3}$$

The denominator is bigger than the numerator so this number must represent a quantity smaller than one.

A number that can be expressed as the division of one whole number by another is called a *rational number*. Notice that this definition of rational numbers includes numbers that are not whole numbers. The number four is a rational number because it can be expressed in a rational form for example

$$4 = \frac{8}{2}$$

But also the number $\frac{2}{8}$ is a rational number according to this definition and it is plainly not a whole number but a fraction of one.

To count up rational numbers that are not whole numbers we will have to use quantities smaller than one.

To count the quantity $\frac{1}{2}$ we need at least one quantity smaller than one. We could use tenths or quarters for example. Contrast this with the number 5. To count 5 we don't need a quantity smaller than 1 although we could count 5 with halves etc. But that is not required.

124   CHAPTER 7. FRACTIONS

Exercise 7.2. Why is the quantity $9 + \frac{1}{2}$ *not* a whole number?

Exercise 7.3. Indicate whether the following number is a whole numbers or only rational number and explain your reasoning: $678 + \frac{1}{2}$.

Solution: If we need a quantity that is a merely a part of one to count the number in question, then this number is not a whole number.

Exercise 7.4. Discuss what we mean when we say that if you are only working with addition and multiplication and you start with positive whole numbers you will only produce more whole numbers. Contrast with division.

To illustrate the idea of a part of one, let's look at the fraction $\frac{1}{3}$.

What is one third?

The starting place is one whole. This can be a quantity 1. We might also call this a *unit* which just means 1.

If we break this quantity, 1, into three equal parts the sum of these three parts is of course 1 and each of these parts is called one third. We write each part $\frac{1}{3}$ and we have this crucial relationship:

$$\frac{1}{3} + \frac{1}{3} + \frac{1}{3} = \frac{3}{3} = 1$$

Exercise 7.5. Referring to the discussion of the meaning of one third, what would be the meaning of $\frac{1}{4}$?

Solution: The quantity $\frac{1}{4}$ is the part of 1 such that the sum of four of these parts equals 1:

$$\frac{1}{4} + \frac{1}{4} + \frac{1}{4} + \frac{1}{4} = \frac{4}{4} = 1$$

Exercise 7.6. Similarly, set out the definition for the fractions $\frac{1}{5}, \frac{1}{6}, \frac{1}{7}, \frac{1}{8}, \frac{1}{9}$, and $\frac{1}{10}$.

Exercise 7.7. What is $\frac{5}{3}$?

Solution: It is the number counting out 5 of the quantity $\frac{1}{3}$. If you take $3 \times \frac{1}{3}$ that is 1. So

$$\frac{5}{3} = 1 + \frac{2}{3}$$

7.2   Unit Fractions

A *unit fraction* is a rational number with numerator equal to 1.

$\frac{1}{3}, \frac{1}{10}, \frac{1}{100}$ are unit fractions. You can see that these unit fractions express quantities smaller than one. You can count these quantities and you can easily express that counting with a numerator.

If you count two of the quantity $\frac{1}{3}$, you write $\frac{2}{3}$.

$$\frac{2}{3} = \frac{1}{3} + \frac{1}{3}$$

Notice again that three of the quantity $\frac{1}{3}$ is $\frac{3}{3}$ and that is one (from the definition of division, how many times must you count three to get three? One time).

The idea of unit fraction provides a simple way to think about any fraction. You can just think of that fraction as a counting of the unit fraction specified by the denominator. The quantity counted is indicated by the numerator.

Twelve thirty-ninths, $\frac{12}{39}$ is 12 of the quantity $\frac{1}{39}$.

So now we have two ways to think about a division expression: from the viewpoint of division itself and from the perspective of counting a unit fraction.

$$\frac{5}{12}$$

We can think of this expression as a division by 12.
We can also think of this as a counting of the unit fraction $\frac{1}{12}$.

Example 7.1. Counting five twelfths.

In the perspective of a counting, the numerator of the division tells us how much we are counting of the denomination. The denominator indicates the unit fraction.

Exercise 7.8. What is the unit fraction we are counting in the expression $\frac{2}{3}$?

Solution: The denominator is 3. This division expression is counting the unit fraction $\frac{1}{3}$.

Exercise 7.9. What is the unit fraction we are counting in the expression $\frac{3}{8}$?

Exercise 7.10. What is the unit fraction we are counting in the expression $\frac{50}{100}$?

Exercise 7.11. What is the unit fraction we are counting in the expression $\frac{12}{1,000,000}$?

Exercise 7.12. Identify the unit fraction being counted and the quantity of this unit fraction counted in this expression: $\frac{3}{10}$.

Solution:

$$\frac{3}{10} = 3 \cdot \frac{1}{10}$$

The unit fraction we are counting is $\frac{1}{10}$. The expression counts 3 of these unit fractions.

CHAPTER 7. FRACTIONS

Exercise 7.13. Identify the unit fraction being counted and the quantity of this unit fraction counted in this expression: $\frac{13}{10}$.

Exercise 7.14. Identify the unit fraction being counted and the quantity of this unit fraction counted in this expression: $\frac{3}{4}$.

Exercise 7.15. Identify the unit fraction being counted and the quantity of this unit fraction counted in this expression: $\frac{4}{3}$.

Exercise 7.16. Identify the unit fraction being counted and the quantity of this unit fraction counted in this expression: $\frac{3}{100}$.

Exercise 7.17. Identify the unit fraction being counted and the quantity of this unit fraction counted in this expression: $\frac{100}{3}$.

Exercise 7.18. Identify the unit fraction being counted and the quantity of this unit fraction counted in this expression: $\frac{3}{1000}$.

Exercise 7.19. Identify the unit fraction being counted and the quantity of this unit fraction counted in this expression: $\frac{300}{1000}$.

Exercise 7.20. Identify the unit fraction being counted and the quantity of this unit fraction counted in this expression: $\frac{3000}{1000}$.

Exercise 7.21. Identify the unit fraction being counted and the quantity of this unit fraction counted in this expression: $\frac{3000}{10}$.

Exercise 7.22. Identify the unit fraction being counted and the quantity of this unit fraction counted in this expression: $\frac{7}{8}$.

Exercise 7.23. Identify the unit fraction being counted and the quantity of this unit fraction counted in this expression: $\frac{14}{16}$.

Exercise 7.24. Identify the unit fraction being counted and the quantity of this unit fraction counted in this expression: $\frac{28}{32}$.

Another way to think of unit fractions is to approach them from the perspective of the denominator.

The denominator determines how many equal parts we want to divide one. Figure 7.1 illustrates dividing a pizza into 12 pieces each piece is $\frac{1}{12}$ of the pizza.

The bigger the denominator, the smaller the unit fraction. This is illustrated in figure 7.2 showing the same circle divided into quarters, eighths, and sixteenths.

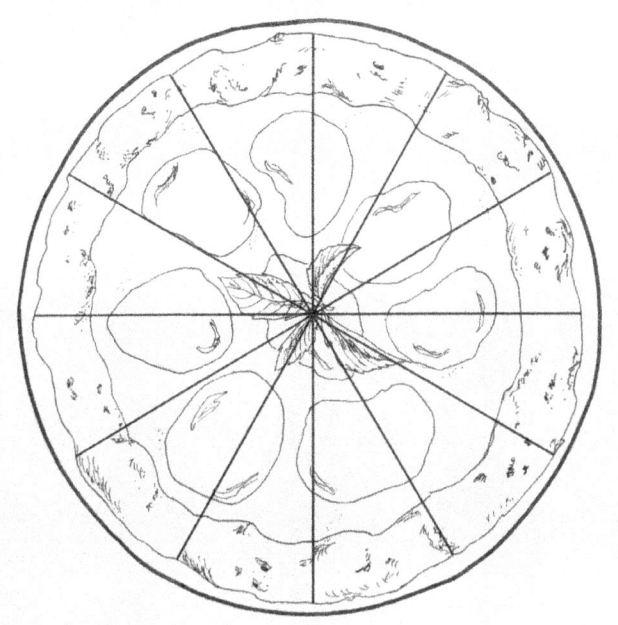

Figure 7.1. A pizza divided into twelve pieces.

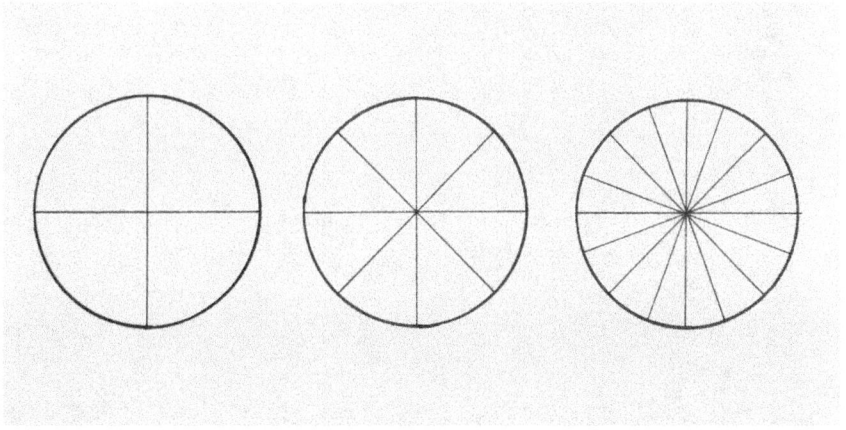

Figure 7.2. Notice that the more equal-sized pieces that we fit in the circle, the smaller must be each piece.

Exercise 7.25. Which quantity is bigger?
$$\frac{3}{11} \quad \text{or} \quad \frac{3}{10}$$

Exercise 7.26. Which quantity is bigger?
$$\frac{1}{10} \quad \text{or} \quad \frac{1}{100}$$

Exercise 7.27. Which quantity is bigger?
$$\frac{1}{10} \quad \text{or} \quad \frac{3}{100}$$

Exercise 7.28. Which quantity is bigger?
$$\frac{7}{8} \quad \text{or} \quad \frac{7}{16}$$

Exercise 7.29. Which quantity is bigger?
$$\frac{3}{4} \quad \text{or} \quad \frac{3}{8}$$

Exercise 7.30. Which quantity is bigger?
$$\frac{2}{4} \quad \text{or} \quad \frac{4}{8}$$

Exercise 7.31. Which quantity is bigger?
$$\frac{8}{1} \quad \text{or} \quad \frac{8}{8}$$

Exercise 7.32. Which quantity is bigger?
$$\frac{9}{10} \quad \text{or} \quad \frac{99}{100}$$

Exercise 7.33. Which quantity is bigger?
$$\frac{16}{1} \quad \text{or} \quad \frac{16}{16}$$

Exercise 7.34. Which quantity is bigger?
$$\frac{16}{1} \quad \text{or} \quad \frac{32}{16}$$

Exercise 7.35. Which quantity is bigger?
$$\frac{5}{25} \quad \text{or} \quad \frac{2}{5}$$

7.2.1 Approaching Zero with Unit Fractions

If we divide 1 by bigger and bigger numbers, we produce smaller and smaller quantities. This is a way to devise numbers that get closer and closer to zero. Let's consider several unit fractions.

$$\frac{1}{1}, \frac{1}{2}, \frac{1}{10}, \frac{1}{12}, \frac{1}{1000}$$

Each unit fraction here is closer to zero than the prior one. And we can see in this method of simply writing unit fractions with bigger denominators that we can always find a number closer to zero. We just choose a bigger denominator. For example,

$$\frac{1}{1,000,000}$$

Exercise 7.36. How much closer to zero is $\frac{1}{3}$ compared to $\frac{1}{2}$?

Draw this out on a line with points showing zero, one third, and one half.

Solution: We determine the distance that we move closer to zero, this is the difference between $\frac{1}{2}$ and $\frac{1}{3}$.

$$\frac{1}{2} - \frac{1}{3} = \frac{3-2}{6} = \frac{1}{6}$$

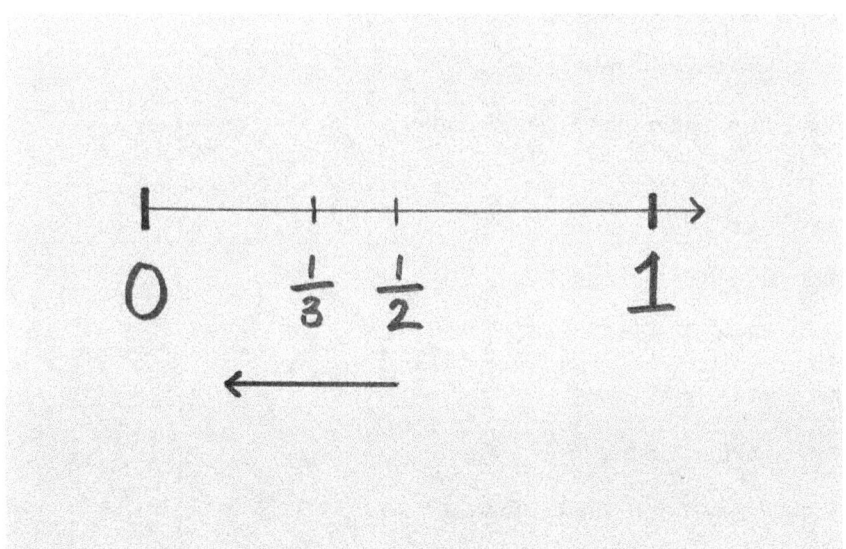

Figure 7.3. A number line with zero, one third, and one half.

Figure 7.4 illustrates how unit fractions with successively bigger denominators get closer to zero.

130   CHAPTER 7. FRACTIONS

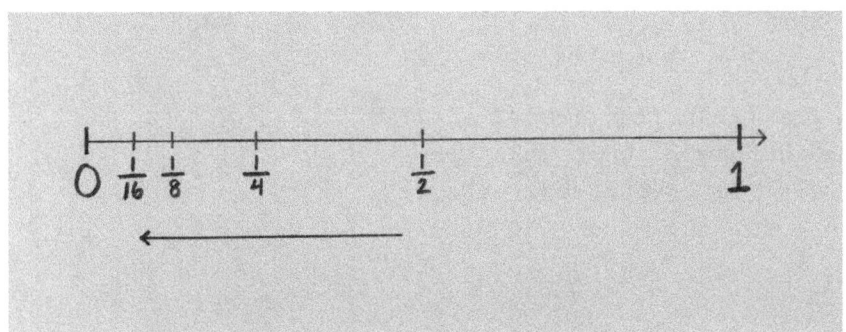

Figure 7.4. Unit fractions approaching zero.

Notice now that unit fractions give us a simple step by step method for approaching zero. We have a way to approach zero (closer and closer) with unit fractions now as well as a way to count towards infinity with whole numbers. *We can always find a number closer to zero than any given small number. And we can always find a number bigger than any given big number.*

Exercise 7.37. What is a number closer to zero than the number $\frac{1}{1,000}$?

Solution: Any unit fraction with a denominator bigger than $1,000$ is smaller than $\frac{1}{1000}$. For example, $\frac{1}{10,000}$. Or also $\frac{1}{1,001}$.

Exercise 7.38. What is a number bigger than $1,000$?

Exercise 7.39. What is a number closer to zero than the number $\frac{2}{1000}$?

Exercise 7.40. What is a number closer to zero than the number $\frac{10}{1000}$?

Exercise 7.41. What is a number bigger than $1,043,599$?

Exercise 7.42. What is a number smaller than the number $\frac{1}{6}$?

Exercise 7.43. What is the biggest fractional power of 10 smaller than $\frac{1}{12}$?

Solution: A fractional power of ten is a unit fraction with a power of ten in the denominator. Ten is smaller than 12 so one tenth is too big. The next fractional power of ten is one hundredth. So the answer here is $\frac{1}{100}$.

Exercise 7.44. What is the biggest fractional power of 10 smaller than $\frac{1}{50}$?

Exercise 7.45. What is the biggest fractional power of 10 smaller than $\frac{2}{3}$?

Exercise 7.46. Describe what is $\frac{1}{13}$?

Solution: This is a unit fraction such that if we take thirteen copies of it their sum is 1.

Exercise 7.47. Similarly, describe what is $\frac{1}{2}$?

7.2.2 Counting Unit Fractions

If you are comfortable with unit fractions then you will be able to work with any fraction because any fraction can be expressed as a counting of a unit fraction.

$$\frac{2}{5} = \frac{1}{5} + \frac{1}{5} = 2 \cdot \frac{1}{5}$$

The numerator 2 in the fraction $\frac{2}{5}$ indicates we are counting 2 of the unit fraction $\frac{1}{5}$.

We can show this geometrically measuring on a number line. Fig 7.5 shows $\frac{2}{5}$ measured out on a line.

Figure 7.5. Two fifths marked out on a number line.

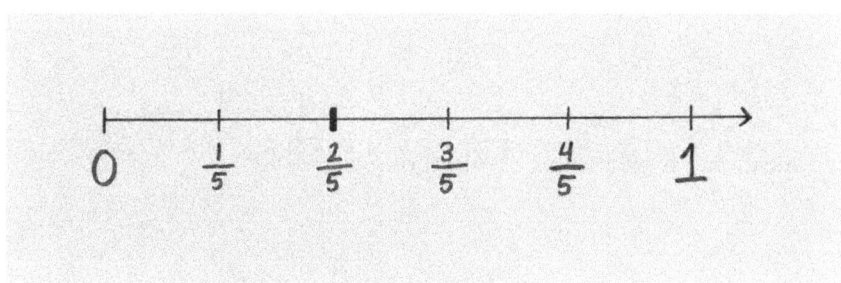

Here are two other representations of fractional quantities. Figure 7.6 shows a square and a circle divided into quarters.

Exercise 7.48. Draw a line. Figure out how to cut it in thirds, fourths, and fifths.

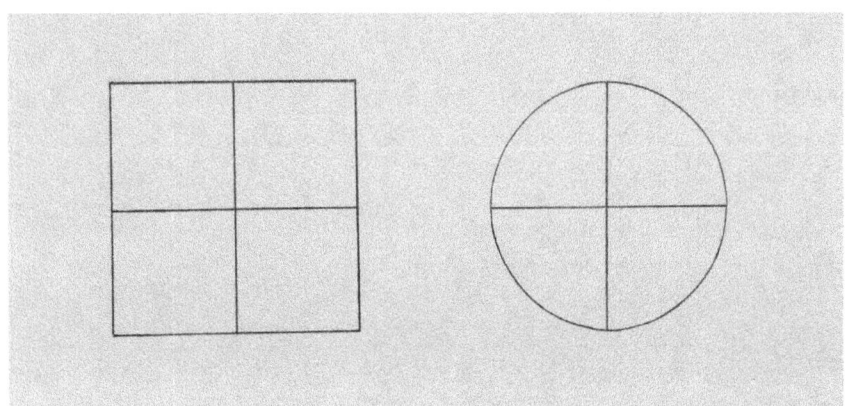

Figure 7.6. Quarters in a square. Quarters in a circle.

7.2.3 One as a Sum of Fractions

Notice that 1 is the sum of unit fractions. The denominator indicates how many terms are in the sum.

$\frac{1}{2} + \frac{1}{2}$
$\frac{1}{3} + \frac{1}{3} + \frac{1}{3}$
$\frac{1}{4} + \frac{1}{4} + \frac{1}{4} + \frac{1}{4}$
etc.

We can express this also using multiplication. This shows us clearly the idea of counting unit fractions:

$1 = 2 \cdot \frac{1}{2}$
$1 = 3 \cdot \frac{1}{3}$
$1 = 4 \cdot \frac{1}{4}$
etc.

Exercise 7.49. What quantity should we add to the fraction $\frac{3}{4}$ to get 1?

Exercise 7.50. What quantity should we add to the fraction $\frac{2}{5}$ to get 1?

Exercise 7.51. What quantity should we add to the fraction $\frac{8}{10}$ to get 1?

Exercise 7.52. What quantity should we add to the fraction $\frac{70}{100}$ to get 1?

Exercise 7.53. What quantity should we add to the fraction $\frac{3}{6}$ to get 1?

Exercise 7.54. What quantity should we add to the fraction $\frac{1}{2}$ to get 1?

Exercise 7.55. What quantity should we add to the fraction $\frac{4}{8}$ to get 1?

Exercise 7.56. What quantity should we add to the fraction $\frac{5}{9}$ to get 1?

Exercise 7.57. What quantity should we add to the fraction $\frac{1}{4}$ to get 1?

Exercise 7.58. What quantity should we add to the fraction $\frac{3}{5}$ to get 1?

Exercise 7.59. What quantity should we add to the fraction $\frac{6}{7}$ to get 1?

Exercise 7.60. What quantity should we add to the fraction $\frac{1}{12}$ to get 1?

Exercise 7.61. What quantity should we add to the fraction $\frac{1}{100}$ to get 1?

Exercise 7.62. What quantity should we add to the fraction $\frac{90}{100}$ to get 1?

Exercise 7.63. Which fractional quantity is bigger, $\frac{1}{10}$ or $\frac{1}{100}$? What does the numerator tell you? What does the denominator tell you?

Exercise 7.64. Which fractional quantity is bigger, $\frac{5}{10}$ or $\frac{49}{100}$? Why? What does the numerator tell you? What does the denominator tell you?

Exercise 7.65. Which fractional quantity is bigger, $\frac{6}{20}$ or $\frac{59}{200}$? Why? What does the numerator tell you? What does the denominator tell you?

7.2.4 Rational Numbers

In the general form of a division of integers n and d, we call n the *numerator* and d the *denominator*:

$$\frac{n}{d}$$

We designate numbers that can be put in this form as *rational numbers*. Although sometimes we might use informally the term *fraction* or *fractional form* to mean also a rational number in the sense that the numerator might be bigger than the denominator.

$$\frac{24}{3} = 8$$

Example 7.2. The numerator is 24. The denominator is 3. $\frac{24}{3}$ is a rational number and also a whole number because exactly 8 threes count 24.

As we have seen, we can think of division as a counting. The denominator tells us what is the quantity to count. Literally this is the denomination. In $\frac{24}{3}$ the denominator is 3. So this division indicates that we count threes.

The numerator tells us how much to count, how much to numerate. In $\frac{24}{3}$ the numerator is 24 so we count the denominator, 3, up to 24 and not beyond. So this division expresses the number of threes needed to count 24.

Notice here that the numerator is greater than the denominator so it is easy to think of counting up the denominator until we reach the numerator.

Let's look at a division where the numerator is smaller than the denominator. Say one divided by five, $\frac{1}{5}$.

How many times must we count 5 to get 1?

We cannot count even a single whole 5 to get one because 5 is bigger than one. So even just a single five is too big of a quantity to express the quantity one. We need only a part of 5 to count 1. If we divide 1 into five equal parts then all five of them together add to 1.

$$\frac{1}{5} + \frac{1}{5} + \frac{1}{5} + \frac{1}{5} + \frac{1}{5} = 1$$

So a single of these parts, that is one fifth or $\frac{1}{5}$, is precisely the fractional quantity that is five times smaller than one.

$$5 \cdot \frac{1}{5} = 1$$

If we are counting in reference to five, to use the approach described above, and we want to count just the quantity one, then we must count the part $\frac{1}{5}$ of 5. One fifth is the part of five that we need to get 1.

We will look carefully later on at how to represent fractions in decimal form—that is, how to count up fractions like $\frac{1}{5}$ using the (fractional) powers of ten. That is a question of representation. How to represent this quantity one fifth.

Here is another way to think about divisions where the numerator is smaller than the denominator.

In $\frac{1}{5}$, the denominator, 5, tells us that we need 5 of the quantity $\frac{1}{5}$ to have a whole 1.

If we have, say, 3 of this quantity $\frac{1}{5}$, then you see we still have just a fraction of one. One is $5 \cdot \frac{1}{5}$ or five fifths whereas we have only $\frac{3}{5}$ (three fifths).

$$\frac{3}{5} = \frac{1}{5} + \frac{1}{5} + \frac{1}{5}$$

$\frac{3}{5}$ is still two fifths less than 1.

$$\frac{3}{5} + \frac{2}{5} = 1$$

Also we can think of three fifths as a counting of $\frac{1}{5}$ like this:

$$\frac{3}{5} = 3 \cdot \frac{1}{5}$$

In this notation we see clearly the numerator as indicating a counting also. The numerator tells us *how many* fifths to consider. The number three fifths tells us that we are considering the quantity that is three times the quantity one fifth.

So we can think of a fraction in two equivalent ways. First by focusing on the quantity expressed in the division. For example $\frac{3}{5}$ expresses a quantity, three fifths, which is a part of one.

Or, second, we can focus on the numerator n and the denominator d and think of the numerator as counting up the fraction $\frac{1}{d}$. In this case we are thinking of $\frac{3}{5}$ as $3 \cdot \frac{1}{5}$.

7.2.5 A Geometric Representation of Unit Fractions

As we have seen, a unit fraction is also a division of one. For example $\frac{1}{8}$ means divide the quantity one into eight equal parts. So far we have tried to express this idea with numbers. And we can also think about it geometrically.

For example we can take a length, say 10 inches, and we can divide that length into eight equal parts. This representation of the division of one lets us visualize the quantity $\frac{1}{8}$ in relation to a specified unit. See figure 7.7.

Figure 7.7. Eighths on a line.

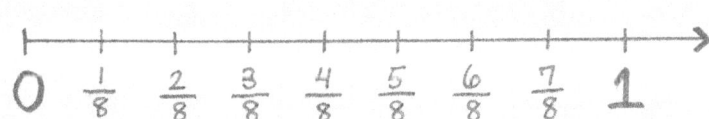

Exercise 7.66. What part of one foot is one inch?

Exercise 7.67. What part of one yard is one foot?

Exercise 7.68. What part of one mile is one yard? Likewise, what part of one kilometer is one meter?

Exercise 7.69. What part of one dozen are three eggs?

7.3 Whole & Fractional Quantities

We have seen that it is often helpful to think of the arithmetic operations as different ways to express and to count quantities. All of arithmetic can be approached in this perspective, which we can extend to fractions as we have also seen.

For example, if we have in mind the quantity five, we might express it as a sum or as a multiplication or as a division:

$$5 = 2 + 3 = 5 \cdot 1 = \frac{5}{1} = \frac{25}{5}$$

Until our discussion on division all of the quantities that we had considered could be counted with one. For example 5 can be expressed as the sum of five ones. Negative eight is $(-1) \cdot 8$. The quantities we were counting did not require a part of one in the way that we need a half or a counting of a half if we want to count up three and a half. Similarly, if we wish to count parts of one like a quarter or an eighth. And we will work with quantities like this all the time. To express these kinds of quantities we must use parts of one. These kinds of quantities we call *fractions*. Division provides a convenient way to express fractions.

Exercise 7.70. We have one cake for three people. We want to divide the cake into three equal parts. How much cake should each person get? How do we express it?

Solution: We divide the cake into three equal parts. Each person gets one part.

$$\frac{1}{3}$$

This quantity, one third, is smaller than one. In fact

$$1 = 3 \cdot \frac{1}{3}$$

We can also write this

$$\frac{1}{3} + \frac{1}{3} + \frac{1}{3} = 1$$

One third is thus a part of one. It is a fraction. It is the part we get from breaking one into three equal sized quantities.

More generally we refer to divisions of the form

$$\frac{n}{d}$$

as fractions where n and d are integers.

For example in
$$\frac{10}{2} = 5$$
the division (10 ÷ 2) defines a fraction of ten. Five is a fraction of ten, it is one $\frac{1}{2}$ of ten.
$$5 = \frac{1}{2} \cdot 10$$
Notice that a whole number can also be expressed as a division, that is in a fractional form.

$$10 = \frac{50}{5} = \frac{1}{5} \cdot 50$$

Example 7.3. 10 can be expressed as a fraction of 50.

There are many ways to express 10 as a division. One such way is $\frac{50}{5}$. This representation highlights that the quantity 10 is one fifth of fifty.

Exercise 7.71. What is ten divided by five?

Solution: We could just treat this using our first definition of division, here: how many times must we count 5 to get 10?

But another way to think about this as a counting of the quantity $\frac{1}{5}$.
$$\frac{10}{5} = \frac{5}{5} + \frac{5}{5} = 1 + 1 = 2$$

Exercise 7.72. Compute these divisions using the approach of counting the unit fraction indicated by the denominator.
$$\frac{40}{8} = ?$$

Exercise 7.73.
$$\frac{36}{9} = ?$$

Exercise 7.74.
$$\frac{15}{5} = ?$$

Exercise 7.75.
$$\frac{70}{10} = ?$$

Exercise 7.76.
$$\frac{24}{8} = ?$$

Exercise 7.77.
$$\frac{36}{18} = ?$$

Exercise 7.78.
$$\frac{60}{5} = ?$$

Exercise 7.79.
$$\frac{56}{8} = ?$$

7.3.1 Expressing a Whole Number in Fractional Form

Any whole number, n, can be expressed in fractional form.

$$n = \frac{n}{1}$$

Recall from division that we can think of this as *how many times must we count 1 to get n?* n times.

$$8 = \frac{8}{1}$$

Example 7.4. Eight as a counting of ones.

Recall also that for any whole number m we have

$$\frac{m}{m} = 1$$

$$\frac{3}{3} = 1$$

Example 7.5. Three is one count of a three.

We see with these relationships that there are many ways to express a whole number in fractional form.

$$1 \cdot 8 = \frac{3}{3} \cdot \frac{8}{1} = \frac{3 \cdot 8}{3 \cdot 1} = \frac{24}{3}$$

Example 7.6. Expressing 8 in thirds.

Exercise 7.80. Express 100 using the quantity $\frac{1}{2}$.

Solution: We want to count 100 with halves. What would be our counting if we were counting with ones? $100 = 100 \cdot 1$.

We also know that $2 \cdot \frac{1}{2} = 1$. There are two halves in 1. This means if we are counting 100 ones, then we are counting $2 \cdot 100 = 200$ halves.

$$100 = 100 \cdot 1 = 100 \cdot (2 \cdot \frac{1}{2}) = (100 \cdot 2) \cdot \frac{1}{2} = 200 \cdot \frac{1}{2}$$

Exercise 7.81. Express 50 using the quantity $\frac{1}{4}$.

Exercise 7.82. Express 25 using the quantity $\frac{1}{2}$.

Exercise 7.83. Express 25 using the quantity $\frac{1}{3}$.

Exercise 7.84. Express 10 using the quantity $\frac{1}{16}$.

Exercise 7.85. Express 11 using the quantity $\frac{1}{8}$.

Exercise 7.86. Compute $\frac{3}{4}$ of 16.

Exercise 7.87. Compute $\frac{1}{5}$ of 100.

Exercise 7.88. Compute $\frac{3}{8}$ of 24.

Exercise 7.89. Compute $\frac{1}{10}$ of 20.

Exercise 7.90. Choose five whole numbers and express each as a fraction of another whole number.

7.4 *Multiplying Fractions*

7.4.1 *A Fraction Times a Whole Number*

Keeping in mind multiplication as a counting, multiplying fractions is simple. For example

$$3 \cdot \frac{2}{5}$$

This is three *times* two fifths. We simply multiply the numerator 2 with 3. The meaning of the expression tells us what we must do. This multiplication is indicating that we take the fraction $\frac{2}{5}$, three times. If we want to express this in a single fraction instead of as a sum of three fractions, then the numerator in our single fraction must be $3 \cdot 2 = 6$ to reflect the correct counting: we now have six fifths.

$$3 \cdot \frac{2}{5} = \frac{2}{5} + \frac{2}{5} + \frac{2}{5} = \frac{6}{5}$$

To multiply a fraction with a whole number we simply multiply the whole number with the numerator in the fraction. The same technique works also for rational numbers more generally. Notice that if the multiplication involves a fraction in the precise sense of a value between zero and 1, then the multiplication acts on the whole number to shrink it. For example if we multiply a whole number by one half, then the product is half the size of the original whole number. Keep this in mind as a quick way to check results.

$$\frac{3}{4} \cdot 12 = 9$$

Example 7.7. The product of $\frac{3}{4} \cdot 12$ can be thought of as shrinking 12 to 9 because it means taking just three fourths—a fraction—of the original whole number, 12.

Exercise 7.91. Compute $2 \cdot \frac{1}{2}$

Solution:

$$\begin{aligned} 2 \cdot \frac{1}{2} &= \frac{2 \cdot 1}{2} \\ &= \frac{2}{2} \\ &= 1 \end{aligned}$$

Exercise 7.92. Compute $3 \cdot \frac{1}{3}$

Exercise 7.93. Compute $4 \cdot \frac{1}{4}$

Exercise 7.94. Compute $5 \cdot \frac{1}{5}$

Exercise 7.95. Compute $6 \cdot \frac{1}{6}$

Exercise 7.96. Compute $7 \cdot \frac{1}{7}$

Exercise 7.97. Compute $8 \cdot \frac{1}{8}$

Exercise 7.98. Compute $9 \cdot \frac{1}{9}$

Exercise 7.99. Compute $10 \cdot \frac{1}{10}$

Exercise 7.100. Compute $18 \cdot \frac{1}{18}$

Exercise 7.101. Compute $49 \cdot \frac{1}{49}$

Exercise 7.102. Compute $100 \cdot \frac{1}{100}$

Exercise 7.103. Compute $1000 \cdot \frac{1}{1000}$

Exercise 7.104. Compute $6 \cdot \frac{2}{12}$

Exercise 7.105. Compute $5 \cdot \frac{4}{20}$

Exercise 7.106. Compute $2 \cdot \frac{1}{2}$

Exercise 7.107. Compute $7 \cdot \frac{3}{21}$

Exercise 7.108. Compute $8 \cdot \frac{7}{56}$

$$\frac{2020}{2020} = 1$$

Example 7.8. There are infinitely many ways to express the value 1 using division. For example $\frac{2020}{2020} = 1$.

Notice that it follows easily from the definition of division that a rational number with the numerator equal to the denominator has value 1. This is very useful to change the denomination of a rational number without changing the overall value as we will see subsequently.

$$1 = 3 \cdot \frac{1}{3} = \frac{3}{3}$$
$$1 = 4 \cdot \frac{1}{4} = \frac{4}{4}$$
$$1 = 5 \cdot \frac{1}{5} = \frac{5}{5}$$
$$1 = 6 \cdot \frac{1}{6} = \frac{6}{6}$$
$$1 = 7 \cdot \frac{1}{7} = \frac{7}{7}$$
$$1 = 8 \cdot \frac{1}{8} = \frac{8}{8}$$

We can write this

$$1 = \frac{n}{n} \quad \text{for any integer } n$$

There are infinitely many ways to express the value 1 using division.

7.4.2 Multiplying Two Fractions

Let's now turn to consider how to multiply two fractions. Say that we would like to compute:

$$\frac{1}{3} \cdot \frac{2}{5}$$

We are here expressing a quantity that is a third of the quantity $\frac{2}{5}$. It designates a quantity that is 3 times smaller than $\frac{2}{5}$. If we multiply the denominator, 5, by 3, we also obtain this quantity.

$$\frac{1}{3} \cdot \frac{2}{5} = \frac{2}{3 \cdot 5} = \frac{2}{15}$$

The quantity $\frac{2}{15}$ is three times smaller than the quantity $\frac{2}{5}$ because the denominator 15 is three times bigger than the denominator 5.

142 CHAPTER 7. FRACTIONS

More generally to multiply two *rational numbers*, we simply multiply their numerators and their denominators. Notice that if both rational numbers are fractions (that is, with values between 0 and 1) then the product of their multiplication will be smaller than each of the original fractions. This is useful for quick checks.

$$\frac{3}{4} \cdot \frac{2}{7} = \frac{3 \cdot 2}{4 \cdot 7} = \frac{6}{28}$$

Example 7.9. We multiply two rational numbers by multiplying numerators and denominators. Keep in mind that making a denominator bigger by a factor c means making the overall value smaller by that factor.

Exercise 7.109. Compute $10 \cdot \frac{1}{2}$

Exercise 7.110. Compute $\frac{3}{4} \cdot \frac{5}{6}$

Solution:
$$\frac{3 \cdot 5}{4 \cdot 6} = \frac{15}{24}$$

We multiply the numerators and the denominators. We can think of multiplication by a rational number in two steps. First, an amplification step corresponding to multiplying the numerator. Here the numerator, 3, amplifies the second number making it three times bigger. Five is thus transformed into 15. The second step is a shrinking corresponding to the multiplication of the denominators. The denominator, 4, shrinks the second number by a factor of four. In the second number instead of dividing by six, we divide by a quantity four times bigger: 24.

Exercise 7.111. Compute and describe this multiplication as a transformation in two steps as indicated in the solution for exercise 7.110.

$$\frac{1}{2} \cdot \frac{1}{3} =$$

Exercise 7.112. Compute and describe this multiplication as a transformation in two steps

$$\frac{1}{4} \cdot \frac{1}{3} =$$

Solution: What does this mean? We are counting a third of one fourth. We cut the quarter into three equal parts and we take one. Note that one third times one fourth is the same as one twelfth.

We can illustrate this with an image. See figure 7.8. Here we have a rectangular grid, three by four. If we first consider this area in thirds and then inside a third we consider the quarters of the third, a single quarter corresponds to a twelfth of the original rectangle.

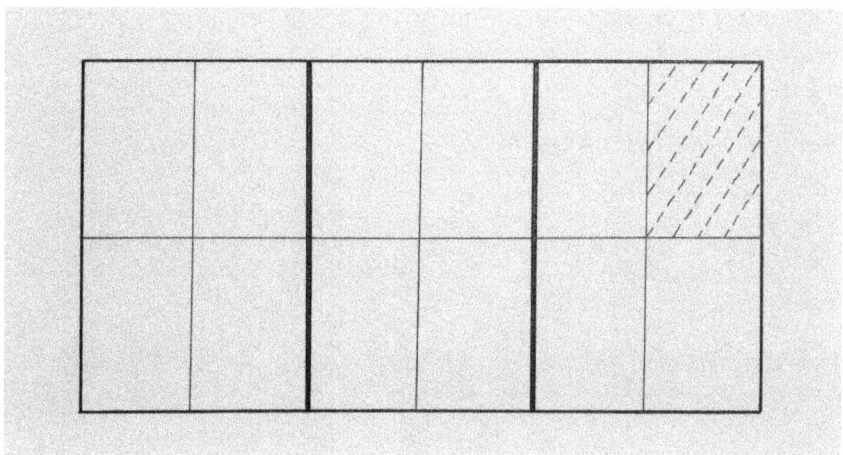

Figure 7.8. If we take a quarter of a third of the rectangle that equals a twelfth of the rectangle.

Exercise 7.113. Compute and describe this multiplication as a transformation in two steps
$$\frac{2}{5} \cdot \frac{2}{7} =$$

Exercise 7.114. Compute and describe this multiplication as a transformation in two steps.
$$\frac{1}{2} \cdot \frac{1}{5} =$$

Exercise 7.115. Compute and describe this multiplication as a transformation in two steps
$$\frac{2}{9} \cdot \frac{1}{3} =$$

Exercise 7.116. Compute and describe this multiplication
$$\frac{3}{14} \cdot \frac{4}{5} =$$

Exercise 7.117. Compute and describe this multiplication
$$\frac{7}{8} \cdot \frac{2}{3} =$$

7.5 Inverses of Fractions

The inverse of a number is a very useful idea and we define it with multiplication. The number m is the inverse of a number n when

$$n \cdot m = 1$$

Notice the symmetry in this relationship. m is also the inverse of n.

$\frac{1}{2}$ is the inverse of 2 because

$$\frac{1}{2} \cdot 2 = 1$$

From this it is also apparent that 2 is the inverse of $\frac{1}{2}$.

Example 7.10. The inverse relationship is very useful and very important.

Exercise 7.118. What is the inverse of $\frac{2}{3}$?

Solution: $\frac{3}{2}$ is the inverse of $\frac{2}{3}$ because

$$\frac{3}{2} \cdot \frac{2}{3} = 1$$

Exercise 7.119. What is the inverse of 5?

Exercise 7.120. What is the inverse of $\frac{1}{10}$?

Exercise 7.121. What is the inverse of $\frac{1}{100}$?

Exercise 7.122. What is the inverse of $\frac{101}{100}$?

Exercise 7.123. What is the inverse of $\frac{8}{7}$?

Exercise 7.124. What is the inverse of $\frac{11}{12}$?

7.6 Dividing Fractions

7.6.1 Division by a Whole Number

To understand division of a fraction by a whole number, focus on the effect of the division. For example:

$$\frac{\left(\frac{2}{3}\right)}{4}$$

We are seeking a value four times smaller than the fraction $\left(\frac{2}{3}\right)$ in the numerator. To make a fraction four times smaller, we can make its denominator four times bigger. We can think of this also as multiplying $\frac{2}{3}$ by $\frac{1}{4}$.

$$\frac{\left(\frac{2}{3}\right)}{4} = \frac{2}{3 \cdot 4} = \frac{1}{6}$$

Because of the relationship between multiplicative inverses, we can always transform a division into a multiplication by the inverse of the denominator. This is often very useful in computing divisions.

$$\frac{\left(\frac{2}{3}\right)}{4} = \frac{2}{3} \cdot \frac{1}{4} = \frac{2}{3 \cdot 4}$$

To divide a fraction by a whole number, n, multiply the denominator by n.

Exercise 7.125.
$$\frac{\left(\frac{5}{8}\right)}{2} =$$

Solution:
$$\frac{\left(\frac{5}{8}\right)}{2} = \frac{5}{8 \cdot 2} = \frac{5}{16}$$

A sixteenth is half of an eighth. So five sixteenths are half of five eighths.

Exercise 7.126.
$$\frac{\left(\frac{3}{10}\right)}{5} =$$

Exercise 7.127.
$$\frac{\left(\frac{3}{10}\right)}{2} =$$

Exercise 7.128.
$$\frac{\left(\frac{3}{100}\right)}{3} =$$

Exercise 7.129.
$$\frac{\left(\frac{3}{100}\right)}{4} =$$

Exercise 7.130.
$$\frac{\left(\frac{4}{5}\right)}{10} =$$

Exercise 7.131.
$$\frac{\left(\frac{4}{5}\right)}{20} =$$

Exercise 7.132.
$$\frac{\left(\frac{5}{16}\right)}{4} =$$

7.6.2 Division of Two Fractions

Consider

$$\frac{\left(\frac{2}{3}\right)}{\left(\frac{1}{4}\right)}$$

If we were dividing $\frac{2}{3}$ by a quantity bigger than 1 we would be making $\frac{2}{3}$ smaller. Here we are dividing by a quantity smaller than one so we are making it bigger. We can compute by using the inverse relationship to change the division into a simple multiplication.

Recall that 4 is the inverse of $\frac{1}{4}$ because

$$\frac{1}{4} \cdot 4 = 1$$

Notice that division by $\frac{1}{4}$ can be expressed as multiplication by 4, the inverse of $\frac{1}{4}$.

$$\frac{n}{\left(\frac{1}{4}\right)} = \left(\frac{1}{\left(\frac{1}{4}\right)}\right) \cdot n$$

In the same way that

$$\frac{n}{x} = \left(\frac{1}{x}\right) \cdot n$$

The inverse of $\frac{1}{4}$ is simply 4, that is:

$$\left(\frac{1}{\left(\frac{1}{4}\right)}\right) = 4$$

This is easy to recall from the inverse relationship:

$$\frac{1}{4} \cdot 4 = 1$$

With this it is simple to re-write a division by $\frac{1}{4}$ as a multiplication by 4:

$$\frac{\left(\frac{2}{3}\right)}{\left(\frac{1}{4}\right)} = \left(\frac{2}{3}\right) \cdot 4 = \frac{8}{3}$$

Notice this product is bigger than $\frac{2}{3}$. Division by a fractional quantity has the effect of magnifying the original quantity.

Another way to think about this division is in referring back to the definition of division. $\frac{n}{d}$ means how many times must we count d to get n. Here the division expresses the idea, how many times must we count $\frac{1}{4}$ to get $\frac{2}{3}$. If we were counting quarters in 2, the counting would be easy: we have four quarters in one so there must be eight quarters in two.

Here we are not counting merely 2 though. We are counting $\frac{2}{3}$. That is a quantity three times smaller than 2. So we will need three times fewer quarters. In other words we won't need the full 8 quarters but rather only $\frac{8}{3}$.

What we did in this perspective is this:

$$\frac{\left(\frac{2}{3}\right)}{\left(\frac{1}{4}\right)} = \frac{\frac{2}{\left(\frac{1}{4}\right)}}{3} = \frac{8}{3}$$

Another way to write this is

$$\frac{\left(\frac{2}{3}\right)}{\left(\frac{1}{4}\right)} = \frac{2}{3} \cdot \frac{1}{\left(\frac{1}{4}\right)} \quad \text{(division as multiplication by a unit fraction)}$$

$$= \frac{2}{3} \cdot \frac{4}{1} \quad \text{(division by unit fraction re-expressed as multiplication by inverse)}$$

$$= \frac{8}{3}$$

To compute divisions of fractions we see it can thus be helpful to write the division as a multiplication by the inverse of the denominator. So with the idea of inverse it is easy to describe how to divide fractions. Simply transform the division into the equivalent expression multiplying the numerator with the inverse of the denominator.

$$\frac{\left(\frac{3}{4}\right)}{\left(\frac{2}{5}\right)} = \frac{3}{4} \cdot \frac{5}{2} = \frac{3 \cdot 5}{4 \cdot 2} = \frac{15}{8}$$

Example 7.11. To divide one fraction by another, simply convert to the equivalent multiplication by the inverse of the denominator.

What we see here is that fractions give us another way to express division. And with the idea of inverse we have a convenient way to compute the division of two fractions.

Note also that now if you know multiplication, you know division.

Exercise 7.133.
$$\frac{\left(\frac{2}{7}\right)}{\left(\frac{7}{8}\right)} =$$

Exercise 7.134.
$$\frac{\left(\frac{3}{4}\right)}{\left(\frac{1}{3}\right)} =$$

Exercise 7.135.
$$\frac{\left(\frac{3}{5}\right)}{\left(\frac{2}{5}\right)} =$$

148 CHAPTER 7. FRACTIONS

Exercise 7.136.
$$\frac{\left(\frac{1}{16}\right)}{\left(\frac{2}{1}\right)} =$$

Exercise 7.137.
$$\frac{\left(\frac{3}{16}\right)}{\left(\frac{4}{1}\right)} =$$

Exercise 7.138.
$$\frac{\left(\frac{21}{7}\right)}{\left(\frac{1}{7}\right)} =$$

Exercise 7.139.
$$\frac{\left(\frac{15}{5}\right)}{\left(\frac{1}{2}\right)} =$$

Let's now take the opportunity of this discussion to practice writing out the division method in a general form. This is an exercise in abstraction. We have two approaches.

Here is a general expression of the division of two rational numbers:

$$\frac{\left(\frac{a}{b}\right)}{\left(\frac{c}{d}\right)} = \left(\frac{a}{b}\right) \cdot \left(\frac{d}{c}\right)$$

Or equivalently we can think of this division in three steps:

$$\frac{\left(\frac{a}{b}\right)}{\left(\frac{c}{d}\right)} = \frac{\left(\frac{a}{b}\right)}{c \cdot \left(\frac{1}{d}\right)} \quad \text{(the denominator as counting a unit fraction)}$$

$$= \frac{\frac{a}{b \cdot c}}{\left(\frac{1}{d}\right)} \quad \text{(dividing the numerator fraction by the first denominator factor)}$$

$$= \frac{a}{b \cdot c} \cdot d \quad \text{(division becomes multiplication by inverse)}$$

$$= \frac{a}{b} \cdot \frac{d}{c}$$

Depending on the numbers one approach might be simpler than the other.

Exercise 7.140. Consider
$$\frac{\left(\frac{3}{4}\right)}{2}$$
Discuss the steps to compute this division.

Solution: First approach. Divide $\frac{3}{4}$ by 2:
$$\frac{\left(\frac{3}{4}\right)}{2} = \frac{3}{8}$$

Second approach. Multiply the numerator's denominator by 2, that is transform the division into a multiplication by the inverse:
$$\frac{\left(\frac{3}{4}\right)}{2} = \frac{3}{4} \cdot \frac{1}{2} = \frac{3}{4 \cdot 2} = \frac{3}{8}$$

The result is the same. In the first approach we take the perspective that this computation entails the division of a counting of the unit fraction $\frac{1}{4}$. So three fourths divided by two is asking us to compute half the amount of the fraction. The division by two is treated here as equivalent to counting the unit fraction $\frac{1}{8}$ instead of $\frac{1}{4}$ because the former is half of the latter.

In the second approach we immediately transform the division by 2 into a multiplication by the inverse.

Exercise 7.141. Consider

$$\frac{3}{\left(\frac{4}{2}\right)}$$ Discuss the steps to compute this division.

Solution: The easy discussion is to transform this expression into a multiplication by the inverse.

$$\frac{3}{\left(\frac{4}{2}\right)} = 3 \cdot \frac{2}{4} = \frac{6}{4}$$

Let's think about this computation also a second way. As two successive divisions of the factors in the denominator. First let's divide the numerator 3, by the 4 (the numerator of the denominator). We can do this easily in thinking of the denominator as four times one half.

$$\frac{3}{\left(\frac{4}{2}\right)} = \frac{3}{4 \cdot \left(\frac{1}{2}\right)} = \frac{\left(\frac{3}{4}\right)}{\left(\frac{1}{2}\right)}$$

Dividing by one half is equivalent to multiplication by 2.

$$\frac{\left(\frac{3}{4}\right)}{\left(\frac{1}{2}\right)} = \frac{6}{4}$$

150 CHAPTER 7. FRACTIONS

Exercise 7.142. Consider

$$\frac{\left(\frac{9}{16}\right)}{3}$$

Discuss the steps to compute this division.

Exercise 7.143. Consider

$$\frac{\left(\frac{14}{32}\right)}{7}$$

Discuss the steps to compute this division.

Exercise 7.144. Consider

$$\frac{\left(\frac{45}{64}\right)}{3}$$

Discuss the steps to compute this division.

Exercise 7.145. Notice again that if we multiply a first quantity by a fractional quantity (smaller than 1), then the product will be smaller than the first quantity. If the first quantity is itself a fraction, then the product will be smaller than each factor. Discuss similarly the cases of dividing a fraction by a quantity bigger than one and also by a fractional quantity (dividing a fraction by a fraction).

7.6.3 A Fraction of a Fraction

It often happens that we are interested in determining a fraction of a quantity. And that quantity can be a whole number or another fraction. This can involve a simple translation into a multiplication expression.

Exercise 7.146. What is a quarter of a dozen?

Solution: One dozen means twelve. So a quarter of a dozen is

$$\frac{1}{4} \cdot 12 = \frac{12}{4} = 3$$

Exercise 7.147. What is a quarter of $\frac{7}{8}$?

Solution: A quarter of seven eighths:

$$\frac{1}{4} \cdot \frac{7}{8} = \frac{1 \cdot 7}{4 \cdot 8} = \frac{7}{32}$$

Exercise 7.148. What is three eighths of two dozen?

Exercise 7.149. What is one quarter of seven miles?

Exercise 7.150. What is three quarters of seven miles?

Exercise 7.151. What is one third of twenty-four hours?

7.7 Simplifying Fractions

We have seen previously that division gives us many ways to express a given value.

$$\frac{1}{3} = \frac{4}{12} = \frac{5}{15} = \frac{6}{18}$$

Example 7.12. Four of the infinitely many fractional forms expressing the value of $\frac{1}{3}$.

It can often be helpful to express a given fractional form more simply or more conveniently in a different denomination. We will see this especially for adding and subtracting fractions.

The key to simplifying a fractional form is recognizing the factors in the numerator and the denominator and the fact that for any number n,

$$\frac{n}{n} = 1$$

This means that we can cancel a factor that appears in the numerator and the denominator without changing the value of the fractional form overall. Cancelling a common factor from the numerator and the denominator is equivalent to shrinking and amplifying the fractional form in the same way.

$$\frac{12}{18} = \frac{6 \cdot 2}{6 \cdot 3} = \frac{6}{6} \cdot \frac{2}{3} = 1 \cdot \frac{2}{3} = \frac{2}{3}$$

Example 7.13. Finding a common factor in 12 and 18.

Twelve and 18 both include the factor 6. In the numerator, this factor is working to amplify the expression by a multiplication of six. In the denominator it is working to shrink the expression by a division by six. These two operations cancel each other. If we remove them both from the expression, we will not change the value of the expression but the numbers remaining are simpler.

Exercise 7.152. Simplify $\frac{12}{16}$

Solution:
$$\frac{12}{16} = \frac{2 \cdot 2 \cdot 3}{2 \cdot 2 \cdot 2 \cdot 2}$$

We see $2 \cdot 2$ in the numerator and also in the denominator.

$$\frac{(2 \cdot 2) \cdot 3}{(2 \cdot 2) \cdot (2 \cdot 2)}$$

We can cancel the common factors $2 \cdot 2$ without changing the value of this quantity because these factors are in both the numerator and the denominator.

152   CHAPTER 7. FRACTIONS

Another way to see this is by noticing that
$$\frac{2 \cdot 2}{2 \cdot 2} = \frac{4}{4} = 1$$
The net effect of these factors in the numerator and the denominator is to multiply the rest of the quantity by one and that does not change that quantity.
$$\frac{12}{16} = \frac{4}{4} \cdot \frac{3}{4} = 1 \cdot \frac{3}{4} = \frac{3}{4}$$
There are other ways to express $\frac{12}{16}$. For example we could have cancelled only the factor 2 in the numerator and the denominator.
$$\frac{12}{16} = \frac{2}{2} \cdot \frac{6}{8} = \frac{6}{8}$$
How much to simplify depends on what will be most useful.

Exercise 7.153. Simplify $\frac{26}{52}$

Exercise 7.154. Simplify $\frac{16}{64}$

Exercise 7.155. Simplify $\frac{12}{60}$

Exercise 7.156. Simplify $\frac{25}{100}$

Exercise 7.157. Simplify $\frac{500}{1000}$

Exercise 7.158. Simplify $\frac{300}{1200}$

7.7.1   Ratio

We have seen that there are many ways to express the same quantity using division. Infinitely many ways. For example, one half
$$\frac{1}{2} = \frac{2}{4} = \frac{3}{6} = \frac{4}{8} \quad \text{and so on}$$
Each of these divisions expresses the quantity one half even though each division itself articulates that quantity using different numerators and denominators.

Similarly with any fraction we can find other divisions (rational numbers or fractional forms) expressing the same quantity but with different numerators and denominators.

Given any fractional quantity, how do we express it with a different division? Very simple, we multiply the numerator and the denominator with the same factor.

Because $\frac{9}{9} = 1$, we do not change the value of the expression although we express it in a different denomination.

$$\frac{2}{3} = \frac{9}{9} \cdot \frac{2}{3} = \frac{18}{27}$$

Example 7.14. To express the fraction $\frac{2}{3}$ in the denomination of $\frac{1}{27}$ we multiply by $\frac{9}{9}$.

If we look more closely we can think of this multiplication in two steps. First a multiplication by nine, then a division by nine. So first we are making the quantity nine times bigger and then we are making that new quantity nine times smaller which brings us back where we started. The net effect of multiplying by $\frac{9}{9}$ is no change. But these multiplications allow us to change the numerator and denominator. In this way, we can express the quantity $\frac{2}{3}$ with a different fraction, $\frac{18}{27}$.

The value itself, which does not change in all these different expressions, is sometimes called the *ratio*. We might speak of the ratio between a numerator and a denominator. As long as we keep the numerator and denominator in the same ratio (or proportion) then the quantity expressed by the division does not change.

$$\frac{3}{4} = \frac{6}{8} = \frac{9}{12} = \frac{12}{16}$$

Example 7.15. These are four different expressions of the same quantity, the ratio of $\frac{3}{4}$.

Exercise 7.159. Simplify $\frac{120}{130}$

Solution: Eliminating from the numerator and denominator the factor in common.

$$\frac{120}{130} = \frac{12 \cdot 10}{13 \cdot 10} = \frac{12}{13}$$

Exercise 7.160. Express $\frac{3}{4}$ in the denomination $\frac{1}{16}$

Solution:

$$\frac{3}{4} \cdot \frac{4}{4} = \frac{3 \cdot 4}{4 \cdot 4} = \frac{12}{16}$$

Exercise 7.161. Express $\frac{7}{8}$ in the denomination $\frac{1}{32}$

Exercise 7.162. Express $\frac{2}{3}$ in the denomination $\frac{1}{24}$

Exercise 7.163. Express $\frac{3}{10}$ in the denomination $\frac{1}{100}$

Exercise 7.164. Express $\frac{3}{10}$ in the denomination $\frac{1}{1000}$

Exercise 7.165. Express $\frac{3}{2}$ in the denomination $\frac{1}{4}$

Exercise 7.166. Express $\frac{15}{3}$ in the denomination $\frac{1}{9}$

Exercise 7.167. What can you say about the value of a rational number if its numerator is smaller than its denominator?

Exercise 7.168. What can you say about the value of a rational number if its denominator is smaller than its numerator?

Exercise 7.169. Express $\frac{3}{7}$ in the denomination $\frac{1}{14}$

Exercise 7.170. Express $\frac{2}{9}$ in the denomination $\frac{1}{90}$

7.8 Adding Fractions

Adding and subtracting fractions is simple once we are comfortable changing the denomination of a fraction. We can think of addition as the combining of a counting. The numerator is counting the quantity indicated by the unit fraction of the denominator.

$$\frac{5}{8} + \frac{3}{4} = \frac{5}{8} + \frac{3 \cdot 2}{4 \cdot 2} = \frac{5}{8} + \frac{6}{8} = \frac{5+6}{8} = \frac{11}{8}$$

Example 7.16. We can think of this addition as the sum of two countings where the first fraction counts the denomination $\frac{1}{8}$ and the second counts the denomination $\frac{1}{4}$. Before we combine these countings we must make sure that we are counting the same thing by expressing both fractions in the same denomination.

7.8.1 Finding A Common Denominator For Two Fractions

Thinking of a fraction as counting the unit fraction indicated in the denominator is a useful approach when searching for a common denominator.

For example the fraction $\frac{5}{8}$ indicates that we are counting the unit fraction $\frac{1}{8}$. The numerator shows the count, five.

Let's consider $\frac{3}{8}$ and $\frac{5}{16}$. The first fraction counts eighths. The second, sixteenths. But one eighth is twice as much as one sixteenth. If we are to combine these two quantities in a single expression, or if we are simply comparing them, we need to account for the difference in the denominators.

We can do this by expressing both fractions in the same denomination.

The easiest situation arises when one fraction can be simplified into the denomination of the other fraction.

We are looking to convert the bigger denominator into the smaller one. Here the bigger denominator is 16. If we divide 16 by 2 we get 8 so it might be possible. Except here in $\frac{5}{16}$ the numerator is 5 and that is not so convenient for factoring 2.

If we had $\frac{3}{8}$ and $\frac{6}{16}$ then we could eliminate the common factor 2 from the numerator and denominator of the second fraction.

$$\frac{6}{16} = \frac{2 \cdot 3}{2 \cdot 8} = \frac{3}{8}$$

A second way to try to bring both fractions to a common denominator is by converting the smaller denominator into the bigger one. Of course we must multiply the corresponding numerator accordingly so as not to change the value of the quantity expressed by the fraction.

$$\frac{3}{8} = \frac{3 \cdot 2}{8 \cdot 2} = \frac{6}{16}$$

If neither of these approaches works conveniently to express the fractions in a common denomination, it is always possible to use the product of the denominators as the common denominator.

$$\frac{1}{3} + \frac{2}{5} = \frac{1 \cdot 5}{3 \cdot 5} + \frac{2 \cdot 3}{5 \cdot 3} = \frac{5+6}{15} = \frac{11}{15}$$

Example 7.17. We can add one third with two fifths by expressing each in the denomination $\frac{1}{15}$.

Exercise 7.171. Compare and discuss: $\frac{11}{21}$ and $\frac{1}{2}$

Exercise 7.172. Choose a few rational numbers and count each with several unit fractions.

Exercise 7.173. Choose a few unit fractions. For each one, practice counting up several rational numbers.

Exercise 7.174. Count with the unit fraction $\frac{1}{3}$, a rational number that cannot be counted with the unit fraction $\frac{1}{2}$.

Exercise 7.175.
$$\frac{3}{4} + \frac{1}{8} =$$

Exercise 7.176.
$$\frac{3}{4} + \frac{4}{8} =$$

Exercise 7.177.
$$\frac{1}{6} + \frac{1}{8} =$$

Exercise 7.178.
$$\frac{3}{6} + \frac{4}{8} =$$

Exercise 7.179.
$$\frac{6}{6} + \frac{7}{8} =$$

Exercise 7.180.
$$\frac{3}{10} + \frac{2}{11} =$$

Exercise 7.181.
$$\frac{3}{10} + \frac{30}{100} =$$

Exercise 7.182.
$$\frac{3}{10} + \frac{2}{20} =$$

Exercise 7.183.
$$\frac{3}{10} + \frac{3}{20} =$$

Exercise 7.184.
$$\frac{9}{10} + \frac{1}{100} =$$

7.8.2 Factorization

From the prior discussion, we see that it is helpful to recognize the factors that compose a number (via multiplication). Let's look at this.

$$12 = 3 \cdot 4$$
$$= 2 \cdot 6$$

Example 7.18. (3, 4) are factors of 12. So are (2, 6).

Exercise 7.185. Find the factors for 60.

Solution:

$$60 = 30 \cdot 2$$
$$= 15 \cdot 2 \cdot 2$$
$$= 3 \cdot 5 \cdot 2 \cdot 2$$
$$= 5 \cdot 3 \cdot 2 \cdot 2$$
$$= 5 \cdot 12$$
$$= 6 \cdot 10$$
$$= 20 \cdot 3$$
$$= 4 \cdot 15$$

One strategy for determining all the factors of a number is to start with a simple division. Then to continue iteratively looking for factors of factors. When you have all prime number factors, then you can start recombining the prime factors to get various arrangement of composed factors. Different factorizations will be useful in different situations.

Exercise 7.186. Find the factors for 20.

Exercise 7.187. Find the factors for 30.

Exercise 7.188. Find the factors for 40.

Exercise 7.189. Find the factors for 64.

Exercise 7.190. Find the factors for 72.

Exercise 7.191. Find the factors for 120.

Exercise 7.192. Find the factors for 52.

Exercise 7.193. Find the factors for 18.

7.8.3 Subtracting Fractions

Subtracting fractions is substantially the same as adding fractions. First ensure each fraction is expressed in the same denomination and then subtract the numerators.

$$\begin{aligned}\frac{3}{5} - \frac{2}{15} &= \frac{3}{5} \cdot \frac{3}{3} - \frac{2}{15} \\ &= \frac{3 \cdot 3}{5 \cdot 3} - \frac{2}{15} \\ &= \frac{9}{15} - \frac{2}{15} \\ &= \frac{9 - 2}{15} \\ &= \frac{7}{15}\end{aligned}$$

Example 7.19. To subtract a fraction from another, first put them in the same denomination and then subtract the numerator.

Exercise 7.194. In $\frac{3}{4} - \frac{2}{3}$ explain why we must first express the fractions in the same denomination prior to subtracting?

Solution: The numerator in each fraction is counting a different quantity. Before we can combine the counts—the numerators—into a single count, the numerators must be counting the same quantity. Otherwise the combination would be meaningless. For example here we might express each fraction in the denomination $\frac{1}{12}$ prior to subtraction.

$$\frac{3}{4} - \frac{2}{3} = \frac{9}{12} - \frac{8}{12} = \frac{1}{12}$$

Exercise 7.195. What might we add to $\frac{5}{8}$ to get 1?

Exercise 7.196. What might we subtract from $\frac{5}{8}$ to get $\frac{1}{2}$?

Exercise 7.197. What might we add to $\frac{5}{8}$ to get 10?

Exercise 7.198. What might we add to $\frac{5}{8}$ to get 5?

158 CHAPTER 7. FRACTIONS

Exercise 7.199. What might we add to $\frac{10}{12}$ to get 2?

Exercise 7.200. What might we subtract from $\frac{10}{12}$ to get $\frac{1}{4}$?

Exercise 7.201. What might we add to $\frac{3}{4}$ to get $\frac{7}{8}$?

Exercise 7.202. What might we subtract from $\frac{3}{4}$ to get $\frac{50}{100}$?

Exercise 7.203. What might we add to $\frac{6}{10}$ to get $\frac{40}{50}$?

Exercise 7.204. What might we subtract from $\frac{6}{10}$ to get $\frac{1}{5}$?

Exercise 7.205. Compute
$$\frac{3}{4} - \frac{1}{5} =$$

Exercise 7.206. Compute
$$\frac{7}{8} - \frac{3}{16} =$$

Exercise 7.207. Compute
$$\frac{2}{9} - \frac{1}{3} =$$

Exercise 7.208. Compute
$$\frac{1}{3} - \frac{2}{9} =$$

Exercise 7.209. Compute
$$200 \cdot \frac{1}{2} =$$

Exercise 7.210. Compute
$$100 \cdot \frac{1}{4} =$$

Exercise 7.211. Compute
$$1000 \cdot \frac{1}{5} =$$

Exercise 7.212. Compute
$$500 - \frac{2,500}{5} =$$

Exercise 7.213. Compute
$$\frac{3}{25} - \frac{4}{5} =$$

Exercise 7.214. Compute
$$\frac{2}{13} - \frac{5}{26} =$$

Exercise 7.215. Compute
$$\frac{2}{13} - \frac{5}{52} =$$

Exercise 7.216. Compute
$$\frac{3}{18} \div \frac{9}{6} =$$

Exercise 7.217. Compute
$$\frac{7}{12} \div \frac{5}{60} =$$

7.9 Proportion

We have a box of 20 pencils. How do we determine a fifth of a box?
$$\frac{1}{5} \cdot 20 = \frac{20}{5} = 4$$

Example 7.20. We multiply the fraction times the whole quantity to determine that fractional part of the whole. We can verify that 4 is one fifth of 20 by noting that $\frac{4}{20} = \frac{1}{5}$.

Exercise 7.218. What is $\frac{7}{9}$ of 54.

Exercise 7.219. What is $\frac{2}{9}$ of 54 knowing $\frac{7}{9}$ of 54.

Exercise 7.220. Discuss. What does *fraction* mean?

Exercise 7.221. What is $\frac{1}{3}$ of 24.

Exercise 7.222. Given that 8 is $\frac{1}{3}$ of 24, use this to explain what quantity corresponds to $\frac{2}{3}$ of 24.

Solution:
$$2 \cdot 8 = 16$$

So 16 is $\frac{2}{3}$ of 24.
We can confirm this noting that
$$\frac{16}{24} = \frac{2 \cdot 8}{3 \cdot 8} = \frac{2}{3}$$

Exercise 7.223. Given that 13 is $\frac{1}{4}$ of 52, use this to explain what quantity corresponds to $\frac{3}{4}$ of 52.

160   CHAPTER 7. FRACTIONS

Exercise 7.224. What is $\frac{2}{3}$ of 27? Discuss.

Exercise 7.225. Given that two thirds of 27 is 18 then what is one third of 27?

Solution:
$$27 - 18 = 9$$
So 9 must be $\frac{1}{3}$ of 27. Check that $\frac{27}{3} = 9$.

Exercise 7.226. What is $\frac{3}{8}$ of 64?

Exercise 7.227. What is $\frac{3}{5}$ of 100?

Exercise 7.228. Let 6 be the quarter of the quantity q. What is q?

Exercise 7.229. What fraction of 6 is 2?

Exercise 7.230. What fraction of 24 is 8?

Exercise 7.231. What fraction of 24 is 6?

Exercise 7.232. What fraction of 60 is 12?

Exercise 7.233. In a scale model airplane that is scaled down by a factor of 48, if the wingspan of the real plane is 50 feet, what is the wingspan of the model?

Exercise 7.234. In a scale model airplane that is scaled down by a factor of 48, if the model propeller has a diameter of 2 inches, what is the propeller diameter of the real plane?

Exercise 7.235. If a coffee costs 50 cents and I want to buy 1,000 coffees, what is the cost?

Exercise 7.236. If a scale model car is two and a half inches long and it is scaled down by a factor of 48, what is the length of the real car?

Exercise 7.237. An easy quantity to relate to is 100. And so it is often used to convey relative size and we speak of *percent* and *percentage*. Percent literally means: for every hundred. For example, eight percent means 8 for every 100. We use the symbol '%' to indicate percent. So we write eight percent also as 8%.

Let's say we learn that about 6% of the Scottish population have red hair. This means that for every 100 people in Scotland, 6 have red hair. If the population of Scotland is 5,500,000, about how many people in Scotland have red hair?

Solution:
$$\frac{6}{100} \cdot 5,500,000 = 6 \cdot 55,000 = 330,000$$
So about 330,000 Scots have red hair.

One way to think about this calculation is that it first batches the total population in groups of 100, since we want to count by one hundreds (for the *percent*). This is achieved via the division by 100. And then, having determined the number of groups of 100, we just count how many red heads there are by multiplying this number of groups by 6 which is the number of red heads we expect in each batch of 100.

Exercise 7.238. If 50% of the cherries in a bag are very sweet and there are 225 cherries in the bag, how many cherries in the bag are very sweet?

Exercise 7.239. If 1% of sea shells on a beach have no chips and we see a patch of 5,000 sea shells, how many have no chips?

www.ingramcontent.com/pod-product-compliance
Lightning Source LLC
Chambersburg PA
CBHW081822300426
44116CB00014B/2457